The Drunken Monkey

The Drunken Monkey

Why We Drink and Abuse Alcohol

Robert Dudley

UNIVERSITY OF CALIFORNIA PRESS
Berkeley   Los Angeles   London

University of California Press, one of the most
distinguished university presses in the United States,
enriches lives around the world by advancing scholarship
in the humanities, social sciences, and natural sciences. Its
activities are supported by the UC Press Foundation and
by philanthropic contributions from individuals and
institutions. For more information, visit www.ucpress.edu.

University of California Press
Berkeley and Los Angeles, California

University of California Press, Ltd.
London, England

© 2014 by The Regents of the University of California

Library of Congress Cataloging-in-Publication Data

Dudley, Robert, 1961–.
   The drunken monkey : why we drink and abuse
alcohol / Robert Dudley.
       pages   cm.
   Includes bibliographical references and index.
   ISBN 978-0-520-27569-0 (cloth : alk. paper)
   ISBN 978-0-520-95817-3 (e-book)
   1. Drinking of alcoholic beverages.   2. Alcohol—
Physiological effect.   3. Alcoholism.   4. Human
evolution.   5. Primates—Evolution.   6. Human
physiology.   7. Monkeys—Physiology.   I. Title.
   GT2884.D84   2014
   394.1'3—dc23   2013033162

Manufactured in the United States of America

23 22 21 20 19 18 17 16 15 14
10 9 8 7 6 5 4 3 2 1

The paper used in this publication meets the minimum
requirements of ANSI/NISO Z39.48–1992 (R 2002)
(*Permanence of Paper*).

To the late Ted Dudley
gentleman, scholar, alcoholic

CONTENTS

List of Illustrations ix
Prologue xi
Acknowledgments xv

1. Introduction 1
2. The Fruits of Fermentation 11
3. On the Inebriation of Elephants 34
4. Aping About in the Forest 51
5. A First-Rate Molecule 69
6. Alcoholics Aren't Anonymous 88
7. Winos in the Mist 115

Postscript 137

Sources and Recommended Reading 141

Index 149

ILLUSTRATIONS

FIGURES

1. Biochemical action of ADH and ALDH enzymes *41*
2. Relative risk of mortality for the fruit fly as a function of exposure to alcohol vapor *45*
3. Relative risk of mortality in relation to alcohol consumption by modern humans *47*
4. Phylogeny of extant apes, with relative extent of frugivory in each group *62*
5. Menu with food and alcohol listings *86*

PLATES

Following p. 48

1. Assortment of rainforest fruits from Barro Colorado Island
2. The palm *Astrocaryum standleyanum* in the rainforest of Barro Colorado Island
3. Fruits of varying ripeness on an infructescence of the rubiaceous shrub *Psychotria limonensis*

4. Extrafloral nectary on a Neotropical shrub
5. A Neotropical fruit-feeding butterfly
6. Fruit flies on naturally fallen figs
7. Ripe fruits of *Astrocaryum standleyanum* on the forest floor
8. An eastern chimpanzee (*Pan troglodytes schweinfurthii*) smelling fig fruit
9. Eastern chimpanzee (*Pan troglodytes schweinfurthii*) and fig fruits
10. Supermarket display of alcoholic beverages
11. The New World phyllostomid great fruit-eating bat (*Artibeus lituratus*)
12. Bonobo (*Pan paniscus*) eating a liana fruit

PROLOGUE

If you walk into any large bookstore and browse in the self-help / recovery section, you will find a number of books about alcoholism. Similarly, a keyword search of books on Amazon will yield in excess of 10,000 items published about the disease. Some are memoirs, others are more clinically oriented, but they will have one major thing in common. All of these books are primarily concerned with the symptoms and management of the disease, rather than with the basic causes of alcoholism. Psychological, sociological, and occasionally physiological underpinnings do receive some attention in these books, but the basic motivation to drink alcohol (either in moderation or to excess) never seems to be explained in detail. Sometimes a spiritual or even a mysterious origin of alcohol attraction is alluded to, rendering any proposed treatment even harder to explain or to interpret from first principles. Most such books would thus seem to be of minimal explanatory or clinical value. However, their very existence and widespread commercial dissemination serve as sad testimony to the hugely detrimental impact of alcoholism, as well as to the desperation of those who suffer from its consequences. Historically, the persistence of alcoholism as a highly damaging medical and sociological phenomenon fully demonstrates our basic lack of understanding as to what might predispose us, as human beings, to suffer from this disease.

My specific interest in alcoholism derives from unfortunate family exposure—my father was an alcoholic who drank heavily, and whose premature death was in part caused by his unsuccessfully treated addiction. Our family, along with tens of millions of other families worldwide, experienced first-hand the sometimes violent and dangerous consequences (including drunk driving) of life with an alcoholic. But perhaps constructively, I well remember as a child being simply puzzled as to why anybody, let alone a parent, might engage in such self-destructive and socially damaging behavior. Although I subsequently pursued research in biomechanics and animal physiology, the answer to this question eluded me until about fifteen years ago, via fortuitous observation of monkeys eating ripe fruit in a rainforest in Central America. Thinking about why the primate brain (or any brain, for that matter) might have evolved the capacity to respond to alcohol, I realized that the taste and odor of the molecule might stimulate modern humans because of our ancient tendencies as primates to seek out and consume ripe, sugar-rich, and alcohol-containing fruits. Alcohol is present because of particular kinds of yeasts that ferment sugars, and this outcome is most common in the tropics, where fruit-eating primates originated and today remain most diverse.

Drawing on my field experiences in China, Malaysia, and Panama, I then developed the idea that fruit consumption by many primates (including our immediate ancestors) prompted the evolution of sensory mechanisms and eating behaviors that are, at least in part, enhanced by the presence of alcohol. This evolutionary outcome would help fruit-eating animals in the wild to rapidly find and consume more calories, and thus to more efficiently feed the hungry primate. I then hypothesized that many if not all of these behaviors, as refined through millions of years of evolution, persist in humans today. Unfortunately, these sensory and dietary responses to alcohol can be co-opted, sometimes for the worse, by the widespread availability and enhanced concentrations of booze present today. What once worked safely and well in the jungle when fruits contained only small amounts of alcohol can be dangerous when we forage in the supermarket for beer, wine, and distilled spirits. As a theory as to why we might be attracted to alcohol, this perspective seemed to have a lot of explanatory power, and also fit well into the emerging field of evolutionary medicine, which emphasizes deep historical roots for many of our current health problems.

In *The Drunken Monkey,* I elaborate on these explanations as to why we drink, sometimes overindulge in, and occasionally abuse alcohol. I particularly seek to provide and to test evolutionary hypotheses for our attraction to beer, wine, distilled alcohol, and other related products of fermentation. When did humans first become attracted to alcohol? Why is it often consumed with food? Why do some people drink to excess? Is there innate genetic protection against alcoholic behavior in certain human groups? And can the study of monkeys and other animals in the wild tell us anything about why and what we drink today? To address these and related questions, I put forward a deep-time and interdisciplinary perspective on modern-day patterns of alcohol consumption and abuse. The sources of information derive from otherwise seemingly unrelated areas of biological knowledge, including how yeasts ferment sugar to produce alcohol, why plants produce fruits, how and why some animals feed on these fruits, and how our drinking behavior today might link with millions of years of evolution within tropical ecosystems. In this book, I develop all of these issues and place them within a unified framework of the comparative biology of alcohol exposure.

Alcoholism, as opposed to the routine and safe consumption of alcohol, remains one of our major public health problems. An important conclusion of *The Drunken Monkey* is that some humans are, in effect, abused by alcohol as it activates ancient neural pathways that were once nutritionally useful but that now falsely signal reward following excessive consumption. Hard-wired responses inherited from our ancestors thus underpin our drinking behavior. This perspective accordingly de-emphasizes the concept of abuse by those addicted to alcohol. Instead, I highlight the biological underpinnings (and associated complexities) of our evolved responses to the molecule. Any approach to understanding contemporary patterns of drinking that fails to incorporate such an evolutionary perspective on human behavior is necessarily incomplete. I have written this book to introduce this new theory of the human-alcohol relationship to the general reader, but also to stimulate further research in this field of scientific inquiry. Alcoholism is a highly damaging disease, both to those who have it and to those who live around them. I can only hope that this book might provide greater insight into its biological and evolutionary origins, and ultimately contribute to its cure.

ACKNOWLEDGMENTS

Much of the "drunken monkey" hypothesis was developed during periods of fieldwork on Barro Colorado Island in the Republic of Panama. I am grateful to the Smithsonian Tropical Research Institute for ongoing support and access to this wonderful field station. Many colleagues have shared their informed opinions about the different ideas presented in this book. I particularly would like to thank Kaoru Kitajima, Doug Levey, and Katie Milton for their critical yet collegial views and overall scholarly assessments of the hypothesis. Carmi Korine and Berry Pinshow had sufficient faith in my early claims about alcohol to begin a collaborative research program on the role of this molecule in the foraging ecology of fruit bats. I still owe them dinner and sake at the finest sushi restaurant in the Negev. At various intervals, Michael Dickinson and Frank Wiens contributed their insights and integrative perspectives on the biology of alcohol consumption. Rauri Bowie, Phyllis Crakow, Phil DeVries, Nate Dominy, Mike Kaspari, Han Lim, Patrick McGovern, Jim McGuire, Sanjay Sane, Bob Srygley, and Steve Yanoviak kindly read the manuscript and constructively pointed out both errors and useful directions for elaboration. Numerous members of my biomechanics research group at Berkeley also provided useful comments on different chapters over many years of manuscript preparation.

My parents-in-law, Mingchun Han and the late Xinping Yan, kindly provided the childcare that enabled completion of the book. I am indebted to Mrs. Rosemary Clarkson of the Darwin Correspondence Project at the Cambridge University Library for providing transcriptions of several unpublished letters by Charles Darwin. These letters, although not proofread to the Project's publication standards, nonetheless yielded wonderful insight into Darwin's views on alcohol as well as his personal drinking habits in his later years. Finally, I thank my wife, Junqiao, my mother, Bettina, and my brother, Topher, for their helpful critique and commentary on the entire text.

CHAPTER ONE

Introduction

Many of us like to drink alcohol, and some of us drink to excess. Why do many people enjoy at most one or two drinks per day, whereas others routinely get plastered? What motivates some college students to drink to the point of passing out or even death? And why do people regularly drink and drive? We have all witnessed examples of both alcohol use and abuse, and perhaps we have wondered why close relatives and friends, when drunk, can behave in aberrant and destructive ways. Alternatively, creative acts of expression and genuine inspiration can result from a glass of wine or a six-pack shared among friends. Where do such differing responses to alcohol come from?

Our relationship with the alcohol molecule is clearly mixed. On the one hand, in social contexts, drinking can be a positive and beneficial experience. Alternatively, it can destroy us, our relatives, friends, and others. And destroy many of us it does, either directly or indirectly. About one-third of highway fatalities in the United States, for example, are alcohol associated. The social, psychological, and emotional damages caused by excessive drinking are more difficult to quantify, but are clearly substantial. Nonetheless, supermarkets, restaurants, bars, and drive-through liquor stores do a thriving business on the sale of alcohol. What factors underlie our drinking behaviors, both responsible and damaging?

This book presents a novel hypothesis to explain our attraction to booze. Unlike many of the addictive substances consumed by modern-day humans, alcohol routinely turns up in natural environments. In the process of fermentation, yeasts that feed on fruit sugars actively produce alcohol, apparently in an effort to kill off competing bacteria that also grow within ripening fruit. Many different kinds of chemical products are generated during this process, but the predominant one is termed ethanol (also known as ethyl alcohol), henceforth referred to simply as alcohol. Not coincidentally, this is the one we prefer. The ecological origin of the alcohol molecule is therefore an important piece of background information if we are to understand our tendency to drink today. Deciphering the origins of fermentation also places them in a much broader ecological context encompassing the biology of yeasts, of microbial competitors such as bacteria, and of the many different kinds of fruit-producing plants.

In the wild, fruits come in all kinds of colors, shapes, sizes, and flavors. And around the world today, there are hundreds of thousands of species of flowering plants, many of which surround their seeds with sweet nutritious pulp. But what makes a fruit ripe and ready to eat, and how do we recognize what constitutes an over-ripe fruit? When might we eat a rotten fruit? At the produce section in the supermarket, we choose fruits on the basis of multiple sensory cues, including their texture, color, and odor. But these products of agricultural domestication differ dramatically from their natural genetic predecessors in the real world. Humans have, via artificial selection over many centuries, created fruits that are typically larger, more sugary, and also more rot-resistant than their wild counterparts. Inferences from our personal experience in the supermarket can therefore be misleading with respect to the natural ecology and ripeness of fruits in nature. To illustrate this point, I'll discuss in chapter 2 the various stages of ripening for wild palm fruits in Panama, starting with their green, unripe, and unpalatable condition, and then progressing to ripe, over-ripe, and finally rotten and disgusting. The ecological microcosm represented by fermenting fruit pulp is a veritable brew of competing viruses, bacteria, and fungi. This point is little appreciated when we consume the banana that was ripe yesterday, but that today tastes a little off.

The high diversity of fruits in nature is paralleled by thousands of different kinds of animals that consume them, including birds (think toucans), mammals (including lots of monkeys and apes), numerous small insect larvae (which we don't really sense or taste when we ingest them), and the ubiquitous microbial community. All of these beasties are competing for access to the sugary nutritional rewards provided by the plant. One ecological definition of ripeness, for example, is suitability for consumption by a vertebrate, mostly birds and mammals, that will consume the fruit and then deposit any ingested seeds somewhere else after passage through the digestive system. Also in chapter 2, I'll describe in detail the evolutionary origins of flowering plants and fruits. Over geological time, mutualistic interaction between animals and the fruits they consume has resulted in greater morphological and physiological diversification in both parties.

Technically, we term the consumption of fruits by animals to be frugivory, and there are many dramatic examples of the extremes to which this evolutionary interaction has proceeded. Consider, for example, the remarkable fishes of the Amazon that travel hundreds of kilometers upriver during the rainy season specifically to eat fruit that has fallen into the waters. Many species in this diverse fish fauna, including the magnificent *piraíba* catfish, which can weigh up to 200 kilograms, engage in this behavior and subsequently relocate the consumed seeds downstream. The local trees of the flooded forests of the Amazon basin are correspondingly specialized to fruit at particular times, so as to facilitate such dispersal. Endless stretches of heavy, fruit-laden branches overhanging riverbanks, and even deeply submerged in water, are an impressive feature of the Amazon and its tributaries during flood season. Ultimately, this spectacle derives from the mutualistic interaction between frugivorous fish and plant.

Other outcomes in this animal-plant relationship are equally interesting. We don't usually think of bears as frugivores, but rather as carnivores. As any reader of the children's classic *Blueberries for Sal* knows, however, at certain times of the year black bears feed almost exclusively on fruit. Similarly, the otherwise terrifying grizzly bear of North America becomes a humble berry specialist as it fattens up for winter in the Rocky Mountains and elsewhere. And how about the toucans, those gaudy birds of Central and South America with wildly enlarged but hollow beaks that are used to manipulate and dehusk fruits plucked from branches high up in the rainforest canopy? Or the enormous fruit bats of the Old World which, as their name indicates, are mostly obligate fruit eaters? These goliaths among the bats, with wingspans up to 1.8 meters, can fly in excess of hundreds of kilometers a night in search of fruit crops, and return to their roosts with their guts laden with pulp and seeds. Another classic fruit-eating mammal is the chimpanzee, our closest living relative, for which over 85% of the diet is typically composed of ripe fruit. In common with many other primates, these animals spend a major fraction of their foraging time traveling to fruit crops and then selecting particular fruits (among thousands in a large tree) for their next meal.

As exemplified by chimpanzees, many of the large fruit-eating mammals are found in lowland tropical rainforests, regions of the world (such as the Amazon and Congo River basins) characterized year-round by high relative humidity and air temperature. Under such conditions, yeasts thrive and ferment. As a consequence, alcohol levels within fruits will be relatively high compared to those in cooler and drier situations in more temperate climates. Animals that routinely eat these fruits for calories, therefore, will also be ingesting alcohol, but the exact amounts and rates of consumption are unknown for any animal in the wild. Among other factors, these will vary with the kind of fruit being consumed, its ripeness and associated internal concentration of alcohol, the regions of the fruit actually being consumed (e.g., the pulp, skin, and seeds), and the total number of fruits eaten per unit of time. Under some conditions, however, enough alcohol may be consumed to result in drunken behaviors that, in humans, we would call inebriation. This outcome has been documented in part by a large popular literature on animal drunkenness in the wild, which is often entertaining but also badly anecdotal (chapter 3). With a few exceptions, the phenomenon of natural inebriation has been little studied. The tendency of some animals to get drunk has nonetheless been known since antiquity. Mythologically, for example, the Chinese monkey king is well-known for a mischievous nature and a taste for alcohol, which yield great confusion and mayhem. Newspapers, and also numerous sources on the internet these days, often report the occurrence of drunken elephants in the Indian subcontinent and of inebriated cedar waxwings in North America. This entertaining and sometimes bizarre literature will be interpreted in chapter 3 from a first-principles scientific perspective.

For at least one group of animals, however, we have some solid evidence as to the behavioral and evolutionary consequences of natural exposure to alcohol. Female fruit flies of many species can smell alcohol vapor emanating from fruits and then fly upwind to find the ripe and over-ripe pulp, upon which they lay their eggs. The larvae then develop in this fermenting mixture and eat not only the sugars but also the yeasts themselves. The alcohol content of the goopy fermented pulp has been well characterized, as have the enzymes within the bodies of the larvae that are involved in the biochemical degradation of the molecule. Two such enzymes are key players in this metabolic pathway, namely alcohol dehydrogenase (abbreviated as ADH) and aldehyde dehydrogenase (ALDH). Genetic variation in ADH and ALDH is widespread in fruit flies and mirrors their natural levels of environmental exposure to alcohol. Fruit flies have long served as a model genetic system in biology, and the study of their responses to alcohol is now yielding insight into the molecular mechanisms of inebriation in humans.

Experimental results with fruit flies will also serve to introduce, in chapter 3, an important physiological concept that is relevant throughout this book. As we will see, remarkable benefits of low-level alcohol exposure accrue both to fruit flies and modern humans. These advantages pertain relative to the conditions of either the complete absence of alcohol or to higher levels of exposure. Such a U-shaped response is a likely evolutionary outcome for many natural substances that occur in the environment at low but persistent concentrations, and which animals may experience on a daily basis. For example, both the longevity and egg output of adult female fruit flies show significant increases following prolonged exposure to low levels of alcohol vapor. Similarly, epidemiological results for humans who drink alcohol at moderate levels suggest surprisingly large health benefits. This outcome is all the more exciting when we consider our very different genetic background relative to that of flies. Nonetheless, the beneficial effects of alcohol make sense when viewed from an evolutionary perspective. As will be seen, so too do the negative consequences of prolonged and excessive drinking associated with human lifestyles in modern environments.

If we then turn to the diet of our forebears among the primates and other mammals (chapter 4), fruit is a routine part of their dinner menu. But ripe fruit is typically hard to find in the tropical rainforest. It can be highly seasonal, and there is ferocious competition among vertebrates, insects, and microbes first to get to the available calories, and then to devour them. A key feature of the drunken monkey hypothesis is that alcohol can be used by all fruit-eating animals as a reliable long-distance indicator of the presence of sugars. As we all know when smelling booze from afar, the alcohol molecule evaporates quickly and can move long distances because of its low molecular weight. And the one commonality of an otherwise bewildering taxonomic and morphological diversity of tropical fruits is that when ripe, they emit an alcoholic signature indicating suitability for consumption. As with fruit flies, any mammal or bird that can sense this signal and then follow it upwind will arrive at the caloric prize. And the quicker the better, so as to eat the fruit before others get there.

Today, lots of insects, birds, and mammals range freely through tropical rainforests doing exactly this, specializing in ripe fruit because of its high caloric returns. Fossil evidence, moreover, indicates that our own primate ancestors were also fruit eaters. Starting about 55 million years ago, primates first turned up on the planet as small tree-dwelling mammals active during the day, probably eating insects. Tens of millions of years later, however, some primate groups switched over to mostly fruits, given what we know from sophisticated anatomical studies of their fossilized teeth. And very suggestively, all of the existing ape species, from gibbons to gorillas, predominantly eat ripe fruit. The only exception to this trend are the highland gorillas, which concentrate on herbaceous and grassy vegetation because large fleshy fruits tend to be absent from the high-elevation flora. Otherwise, down in the steamy humid forests of the lowlands, the apes are happily looking for and consuming squishy ripe fruits most of the time.

Although the great apes (including chimpanzees) are the primates today most closely related to modern humans, the divergence between these two evolutionary lineages actually began close to eight million years ago. The diets of early human ancestors diversified over the following millions of years and began to include a much broader range of plant tissues and greater amounts of animal fats and protein (including, in rude fashion, one another from time to time via cannibalism). The ability to cook both tubers and meat may also have played an important role in dietary outcomes, although the timing of this possibility is hotly disputed. Unambiguously, however, the dinner menu changed dramatically about 12,000 years ago with the origins of agriculture. Cultural evolution in humans then began to exert the predominant influence on what we eat. Nevertheless, as with meat consumption, preference for salt, and a variety of other dietary habits, our eating choices today can be strongly influenced by genetic predisposition. Nowhere is this effect more evident than in the so-called diseases of nutritional excess. These are adverse medical outcomes deriving from a mismatch between the biological environments in which we evolved and the ones that we have created and live in via technology. Alcoholism may be one such disease, as detailed in chapter 4.

Coincident with the development of agriculture, humans innovated the fundamental chemical procedures of brewing, wine-making, and the intentional fermentation of alcoholic beverages (chapter 5). Although these events are impossible to date precisely, chemical analyses of pottery vessels indicate wine-making as early as 7000 BCE. Alcohol production rapidly became an important feature of human social life. Its relevance intensified when improvements in crop productivity and the invention of distillation (probably first in China before 200 CE, but only broadly disseminated by 700 CE) rendered high-concentration alcohol much more available. Industrialization in the nineteenth century further enhanced supply and reduced price. Today, with the notable exceptions of the Islamic world and some major ethnic groups in South and East Asia, the consumption of alcohol forms a major theme in human domestic life. We use alcohol in religious rites, regularly consume it with meals and during celebrations, and often socialize while partially or fully drunk. Restaurants generally derive about half of their profits from the sale of booze. The adverse consequences of excessive alcohol consumption are equally salient—drunken brawls, highway crashes, domestic violence, liver cirrhosis, and premature death. Our attitudes towards this molecule are clearly conflicted. On the one hand, we appreciate the psychoactive and socially relaxing features of a glass of beer or wine. On the other, we can have great difficulty working with the drunkard who also happens to be our essential colleague in the office.

And if the endpoint of extreme drinking can be death for ourselves and possibly others, why then do some of us become irreversibly addicted to alcohol? This critical yet to date unresolved issue is addressed in chapter 6. Part of the problem lies in our genes. Abundant evidence from twins who were separated at birth and subsequently evaluated medically in adulthood demonstrates heritable components to alcoholism. Males are also much more likely to be classified clinically as alcoholics than are females. Nonetheless, the heritability of alcoholism is only partial, and a variety of environmental circumstances as yet not well understood also contribute to the emergence of the disease. The historically diverse array of treatments used (typically without success) to treat alcoholism demonstrates the equally wide range of opinions as to the origins and causes of the syndrome. If we accept, however, an evolutionary and psychoactively rewarding association between alcohol and caloric gain, then a more fundamental explanation becomes clear. What once served us well in searching for and consuming fruit in tropical forests may now be co-opted by the essentially unlimited availability of the alcohol molecule. Intriguingly, an evolutionary signature of those biological underpinnings to addiction is provided by genetically based differences among modern humans in the ability to metabolize alcohol, and correspondingly in their tendencies either to drink excessively or not at all.

One biomedical approach to understanding the effects of alcohol is to study the reactions of other kinds of vertebrates to the molecule. Over the last six decades or so, extensive time and money have been put into the development of various rodent and nonhuman primate systems used to simulate the routine drinking patterns as well as the alcoholic behavior characteristic of humans. As we will also see in chapter 6, this work has mostly failed to yield fundamental insights into the nature of addiction to alcohol. Part of the problem derives from basic biology—the rodent species used in such studies are mostly temperate-zone omnivores, and historically were never much exposed to fermenting fruit. The behavioral and sensory responses of these species to alcohol in laboratory contexts are therefore somewhat abnormal. Equally problematic with these standard mammalian models is the provisioning of liquid alcohol as a supplement to their solid food. This approach may more accurately simulate human drinking patterns, but obviously deviates from the natural commingling of alcohol and nutritious pulp that characterizes fruit fermentation in the wild. By contrast, the study of more realistic and biologically appropriate animal models may help spur understanding of natural behavioral reactions to alcohol, including those addictive responses that characterize our own species.

Although we are entering the new millennium of genomic approaches to human disease, an evolutionary perspective is conspicuously absent from the literature on alcoholism. Instead, current research emphasizes reductionistic and physiological approaches to addictive disorders. A relative novelty for psychoactive compounds is typically assumed in such studies, although the presence of pre-existing neural pathways that underpin addictive behavior must also be recognized. This view is certainly appropriate for most of the chemical substances to which humans nowadays become addicted. Psychoactive and addictive compounds such as nicotine, cocaine, and morphine are found only in a limited number of plant species, and occur at very low concentrations within plant leaves and other structures. Alcohol, by contrast, occupies a unique position in primate nutritional ecology given its obligate and widespread association with fruit sugars. It is therefore fundamentally different from other psychoactive compounds in the extent to which it formed a routine component of our ancestors' diets. Like all species of plants, animals, and fungi, *Homo sapiens* is an evolved outcome. We ignore deeply rooted historical influences on human ecology and biology at our peril. Chapter 7 places the drunken monkey hypothesis for alcoholism within the broader context of evolutionary medicine and suggests a number of directions for future research into this perplexing disease.

In sum, this book seeks to explain the biological underpinnings of our innate attraction to alcohol. It presents an evolutionary interpretation not just for its routine consumption, but also for the excessive use that leads to alcoholism. To reach this point, it is necessary to explore in detail many interesting but seemingly unrelated themes within the biological sciences. These topics include the ecology of tropical rainforests, the biology of fruits and ripening, fermentation by yeasts, animal feeding habits, the history of the human species, contemporary drinking behavior, and the epidemiology of alcoholism. As we will see, these topics and others can all be interpreted within a unified framework of comparative and evolutionary biology. This is a new and challenging perspective on our conflicted relationship with alcohol.

CHAPTER TWO

The Fruits of Fermentation

Rental car companies no doubt have to deal with many kinds of customers and problems, but it was nonetheless surprising, having rented a car at the Kuala Lumpur airport, to read the following notice in our vehicle: "Tariff will be doubled if pungent odor of durian pervades the vehicle." The large and infamous durian fruit of Southeast Asia exudes a powerful smell reminiscent to some of rotting garbage, and to others of sherry trifle. Like so many tropical fruits, the flavor of the pulp is rich and sensuous, albeit with hints of fermentation if not actual decay. It's clearly attractive to lots of animals, and many humans will pay top price for the luscious taste and texture of a ripe durian. How did such botanical exuberance evolve in the first place, and what biological factors have motivated the expression of such tastes and odors? Why should durian pulp be sufficiently smelly as to offend car drivers in Malaysia, and what's the link between fruit sugars and decay?

SWEET AND SQUISHY

It's easy to get lost in a tropical rainforest, because most of it is a lovely green and looks pretty much the same. When we walk about, what we see are mostly leaves, followed by twigs, branches, and tree trunks.

Photosynthesis by leaves, together with the woody structures that support them, represents the primary investment in the economy of plant life. Flowers, fruits, and seeds are much less common in space and in time, but the intermittent bouts of reproductive activity by flowering plants can be spectacular. Floral displays blanket the crowns of trees and shrubs, bright clusters of fruits hang from the ends of branches, and layers of fallen fruit decompose on the forest floor. Most obviously, the often vivid colors of fruits and flowers contrast radically with photosynthetic greenery, indicating different underlying physiologies as well as ecological roles. The typically brief but glorious existence of these reproductive structures provides an esthetic window into the sexual life of plants and into the powerful forces of selection that have molded them through evolutionary time.

Flowering plants, known botanically as angiosperms, originated about 140 million years ago in the geological period termed the Cretaceous. The technical definition of a flowering plant actually refers to the nutritious packaging around the seeds within an angiosperm's fruit, rather than to the flower itself. The associated sugars and fats which nourish the seeds provided a substantial source of energy that was attractive to and readily consumed by rapidly diversifying bird and mammal communities more than one hundred million years ago. As a consequence of eating fruits and then depositing the seeds elsewhere, these animals provided to plants the benefit of long-distance dispersal to new habitats. Since the Cretaceous and onwards into the present, fruits (both dry and fleshy) have become an obvious feature of plant biology in many terrestrial ecosystems. Sugar-rich fruits remain today a common and important aspect of many tropical and temperate-zone forests. When domesticated, these fruits—including such moist and fleshy ones as tomatoes, bananas, and apples, along with dry seed-containing fruits such as grains and nuts—represent a major component of the contemporary human diet. We sensorily experience the evolutionary wonder of fruits every time we wander in the produce section of the supermarket.

The two-way interaction between fruits and their animal dispersers is a well-known example of the evolutionary outcome termed mutualism. Both participants benefit in a mutualistic interaction, and the tightness or specificity of the association often becomes greater with time. Similar dynamics have characterized the evolution of flowers and the various animals that pollinate them in exchange for the caloric reward of sugar-rich nectar. The amazing displays of floral color that we visually appreciate today originated to meet the energy demands of a huge range of insects and vertebrates. Indeed, much of modern-day plant diversity can be linked to the parallel enlistment of floral pollinators and vertebrate consumers attracted for the purposes of nutritional payoff. Ripe fruit thus represents the coinciding interests of animal appetite and the dispersal of plant progeny. One of the consequences of this mutualism has been an increased diversity of morphological, physiological, and (for animals) behavioral traits that facilitate more efficient interspecific interactions. More obvious fruit, better searching strategies, and more specialized vision among frugivores are some of these evolutionary outcomes, along with enhanced species diversity of both angiosperms and animal consumers.

We live in a world dominated ecologically by flowering plants and, by association, their seed-containing fruits. This outcome is most obvious in the tropical and subtropical regions, where riotous assemblages of herbs, shrubs, vines, saplings, and trees vigorously compete for access to light. These forests can be structurally complex, with no well-defined transition between the ground cover and the canopy. Vegetation is simply profuse and confusing to the human eye, yielding the iconic imagery of the tropical rainforest. Also in the tropics, plant species diversity is famously high. Unlike the coniferous forests of boreal regions, tropical rainforests are numerically dominated by flowering plants. Species richness here can be overwhelming even to botanical specialists. For example, Barro Colorado Island in the Republic of Panama has been a nature reserve since the completion of the Panama Canal and the filling of surrounding lowlands with water in 1914. Over

1,250 species of flowering plants can be found on this small island, which is only about sixteen square kilometers in area. By contrast, the more primitive seed-bearing plants that do not bear flowers are represented by only one species. With the exception of the higher-latitude coniferous forests, terrestrial vegetation worldwide is similarly dominated by plants that bear flowers and, in many cases, fleshy fruit. On Barro Colorado Island, the morphological range of fruits produced by the flowering plants is impressive (see plate 1).

Such taxonomic structuring of terrestrial ecosystems was not always the case in earth's history. Prior to the diversification of flowering plants, these ecosystems were dominated by such seed-bearing groups as conifers and cycads, along with numerous lower plants, including tree ferns and mosses. Pollination in these groups is typically by means of wind, or even by water transmission of gametes in more primitive forms. By evolving first flowers and then fruits with internalized seeds (sometimes in the form of nuts), animals could be enlisted in a more targeted dispersal of pollen and fertilized embryos (i.e., seeds). Surrounding the seed with sugars and delicious fats induces consumption of the ripe fruit by vertebrates. Animals seek out these nutritional rewards around the seed and then relocate it elsewhere once it transits through the gut. Sometimes abrasion by the digestive system and its associated enzymes is even a prerequisite for seed germination. During the initial stages of angiosperm evolution, dinosaurs may have consumed their fruits, given that mammals and what we know today as birds (i.e., dinosaurs with wings) were not yet present on the earth. Subsequent evolution of these latter groups in the last sixty million years has been paralleled by corresponding taxonomic and morphological diversity in fleshy and reward-providing fruits.

Nowhere is this biological outcome more apparent than in the tropics. One of the great pleasures on earth is to spend time walking, watching, smelling, touching, and listening within a tropical rainforest. Lush vegetation, bewildering insect species, and hyperdiverse bird and mammal communities compared to those in the temperate zone have alternately inspired and profoundly discouraged scientific investigators. The diversity can simply be overwhelming. And to this day, there is no fundamentally convincing explanation for the so-called latitudinal species gradient, whereby virtually all groups of plants, animals, and fungi are much more species-rich as one approaches the equator. Certainly for sugar-rich fruits and for those animals that consume them, the tropics are home to a number of spectacular evolutionary experiments. For example, street markets in tropical countries typically display a wide range of colorful and fragrant fruits unavailable in the temperate zone. Often intensely sweet, with distinctively aromatic flavors, these fruits, when ripe, are squishy and easily crushed, and therefore are not easily transported to and distributed in more industrialized nations. Such fruit displays reflect more generally the amazing range of plant products available in the forests. In lowland tropical rainforests, anywhere from 50 to 90% of all flowering plants are visited by fruit-eating birds and mammals, which number in the thousands of species worldwide.

One important example of such fruiting trees are the palms. With over 2,600 species found mostly in the tropics, palms provide large quantities of sugary fruits to many different kinds of animals. A typical palm is *Astrocaryum standleyanum*, a common species in lowland Central and South American rainforests. This species bears very large fruit crops (see plate 2), with each cluster weighing up to twenty kilograms. The fruits are consumed by a broad diversity of animals, including red-tailed squirrels, spiny rats, kinkajous (an arboreal carnivore that eats mostly fruit), Central American agoutis (a large rodent), collared peccaries, howler monkeys, and white-faced capuchin monkeys. The palm fruits start out green and unripe but mature over the course of several months to turn a distinctive orange, with sweet, rich, and odoriferous pulp. Some animals manage to surmount the spine-covered trunk of the palm to consume fruit from the heavy clusters. More typically, fruits fall to the ground, where they are stripped of their pulp by various feeders. Agoutis in particular are fond of these palm fruits, and they relocate and bury the seed for future consumption. This is a mutualistic interaction beneficial to both agouti and palm, as not all buried seeds are subsequently found and consumed by the rodent. Those undiscovered will then germinate and contribute to the next generation of palms. Fruits not eaten by animals will rapidly turn a darker orange and then black, becoming truly rotten and disgusting. Bacteria eventually consume all available sugars in this process.

Another major group of fruit providers in the tropics are the figs. With over 750 species in what is the largest plant genus (*Ficus*), big fig trees are a common sight in lowland tropical rainforests. Their ripe fruits provide abundant pickings for some bats, many primates, large birds such as hornbills and toucans, and a diversity of smaller birds and mammals. Figs as well as palm fruits have been termed keystone resources for tropical vertebrate frugivores, providing a substantial fraction of many animals' daily energy requirements. These fruits can be particularly important during periods of scarcity in the forest, when most other plant species are not fruiting because of seasonal weather patterns. Figs and palms, by contrast, provide more reliable crops to the benefit of the animal consumers. Another important feature is that these fruits are often fairly large. In the lowland rainforest of Barro Colorado Island, for example, the average size of figs and palm fruits is about 1.5 centimeters. Such fruits are also typically available in aggregate, often within hanging clusters or bunches, and represent a huge meal to those who can find them.

But before figs can be eaten, they must undergo a complex series of changes to reach the point of being attractive to consumers. First, all fruits must begin their development within a pollinated flower. Following fertilization of the female gametes, the reproductive tissue of a flower grows and begins to sequester nutrients, mostly starch, from other regions of the plant. Seeds within the fruit mature simultaneously but remain inviable up to the point of maturity. The fruit remains green and unpalatable throughout this time, as premature consumption of the fruit by animals would inevitably result in destruction (rather than dispersal) of the seeds. Immature fruits are thus tough and are often chemically defended by nasty tasting compounds (such as tannins) so as to deter such an outcome. Biting into and then spitting out unripe peaches or apples is perhaps the closest we come to experiencing these defenses, but in the real world animals quickly learn to avoid such unripe fruits, except in conditions of extreme hunger. At a certain stage of development, however, fruits become ripe and attractive to their animal consumers. Physical and chemical defenses are relaxed, and otherwise relatively indigestible starch molecules and other complex carbohydrates are converted to simple sugars. The fruit thus becomes sweeter and more attractive to microbes, as well as to monkeys and other vertebrates.

This ripening process involves a number of different internal changes that influence both structural and biochemical properties. In the transition to ripeness (and ultimately to being over-ripe), fruits typically enlarge, increase their water content (effectively becoming juicier), change color, become softer, and reduce their chemical defenses. These features typically change in concert and are regulated by a number of different hormonal pathways. For many fruits consumed by diurnal (i.e., day-active) birds and mammals, the goal of this process is to provide an end product that is both obvious to animals at a distance and ready for consumption. Surface color alone is often sufficient to indicate ripeness, given the marked shift from an unripe green to red, purple, yellow, orange, or even blue in some cases (plate 3). Some fruits exhibit this brightness in the ultraviolet as well. Changes in odor are equally pronounced, as the fruit advertises its presence with a wind-borne aromatic signature composed of many different volatile molecules. Bat-dispersed fruits in particular are characterized by chemical odors signaling ripeness, as visual cues are much less effective for these night-flying animals. Texturally, ripening fruit becomes much softer as cell walls are degraded enzymatically. Sugar content increases dramatically, and the indigestible and sometimes toxic compounds that characterized the green unripe fruit are broken down.

In the tropics, sugars provide the primary nutritional reward within ripe fruits. There are some oily fruits such as avocados, however, that contain little or no sugar. These instead contain calorically dense fats as an enticement for consumption. Oily fruits actually tend to be much more common in temperate-zone ecosystems, where fruits are primarily consumed by birds. In the autumn, migratory birds in particular are major visitors to fruiting trees and shrubs. Fats are much more energetically dense than sugars and are much better suited for long-distance migrants trying to minimize weight. By contrast, ripe tropical fruits contain mostly sugar, with values typically ranging from 5 to 15% of the pulp mass (but occasionally with a sugar content as high as 50%). Tropical fruits also tend to be more watery and larger than those in the temperate zone (think of mangos and papayas, for example). Too much investment in sugar may represent an inordinate cost relative to the potential evolutionary benefits associated with attracting an animal. Nonetheless, such fruits represent a substantial caloric reward to the individual that finds them.

Additional players, however, have contributed to the interesting biological outcome of the ripe fruits that we and other animals enjoy today. Coincident with the evolution of sugar-rich fruits, yeasts evolved the ability to produce alcohol, apparently to kill off bacterial competitors. As green fruits progressively ripen and then rot, various microbial agents of decay grow and develop, simultaneously devouring any available sugars. When a fruit is ripe and ready for consumption, a variety of visual, chemical, and textural cues then advertise the availability of calories to a bird or mammal. We subconsciously use these cues when shopping in the supermarket for fruits and vegetables, and chimpanzees use them high up in rainforest trees when selecting figs for consumption. Where there are ripe tropical fruits containing sugar, there will also be fermentation by yeasts. Those animals that happen to eat these fruits will, therefore, also be inadvertently consuming alcohol. Today's foraging behaviors by birds and mammals are thus superimposed on a historical background of ecological interactions and intense microbial combat within fruits that first turned up millions of years ago. The basic themes of competition for fruit sugars, fermentation, and dietary exposure to alcohol are thus both ancient and persistent.

Our perspectives on fruit are largely shaped in modern industrialized countries by their availability in supermarkets. However, large displays of uniform and unblemished fruit are really not representative of conditions in the wild. Domestication over millennia has yielded major shifts in fruit size (mostly via increased water content), sugar composition, and texture. Most recently, the demands of long-distance transport to market have imposed requirements for durability and ease of packing. Such changes, in aggregate, have produced fruits that but little resemble their natural genetic predecessors. Many fruits in the wild are riddled with insect larvae, fungal rot, and other such infestations. What the grocery industry considers to be ripe fruit ready for consumption is relatively disease-free and sweeter than that typically eaten by animals in the wild. Our concept of ripeness is also mediated culturally. Some people won't peel open a banana if there is a single dark spot on the skin. By contrast, others are much less sensitive to the consumption of over-ripe fruit, particularly when they are hungry.

Ripeness in the real biological world, by contrast, means only that a fruit must be sufficiently adequate for consumption, even if the nutritional rewards are not necessarily ideal from the perspective of the consumer. Selection will only act on a plant's progeny if they survive to reproductive maturity. Any relocation of a seed, no matter how effected, can therefore be advantageous. Fruits are certainly more susceptible to microbial decay once ripeness is attained, given the greater sugar levels. But even during the process of fruit formation and development there exist the possibilities of bacterial infection, germination of yeast spores that landed at the flower stage, mechanical abrasion and consequent microbial invasion while on the twig or branch, and infestation by insects. The relaxation of chemical and structural defenses during ripening inevitably increases the chances of incipient rot and decomposition. Even when ripe, considerable time may pass before a fruit is actually found and eaten, a period during which both bacteria and fungi may flourish. In essence, there exists an ecological race in time between the agents of decomposition and those of consumption. Microbes and animals compete to take advantage of available sugars. If sufficiently far progressed, bacterial rotting and decay can potentially discourage consumption by frugivores. Seeds within the fruit will correspondingly not be relocated far from the parent tree and may suffer increased mortality. Given that microbial growth rates can be really high, particularly in the warm, humid tropics, bacteria and fungi present a real threat to the reproductive interests of flowering plants. Microbes are omnipresent and happily devour plant and animal tissue alike, including our own when given the opportunity.

THE YEASTS OF DECAY

If we observe naturally occurring fruit fall in most regions of North America, individual fruits often stay in place for weeks or even months. Small sections can be consumed by insects and fungal spots may appear, but decomposition proceeds only slowly. Unless a passing animal or human physically removes the fruit, it will remain there, remarkably unperturbed by decay. By contrast, similar observations of fallen fruit in the lowland tropics result in a substantially different outcome. Insects and microbes find and colonize fruit within minutes, and the likelihood of a wild vertebrate removing and eating them is much higher. Decomposition proceeds quickly, and within days the fruit is but a black and rotten remnant. Microbial growth is particularly temperature sensitive, and the elevated and also fairly constant air temperatures in lowland tropical regions predispose fruits to quick decay. Animal carcasses similarly disappear within days in tropical forests, yielding a nasty malodorous plume, along with vultures in abundance.

The theme of rapid decomposition is thus paramount in the humid tropics. Much of this decay occurs internally within the guts of termites, which compose the majority of animal biomass within tropical rainforests. Using protozoans that live in their midguts, termites can successfully degrade the cellulose molecules of plant cell walls. Similarly, varied kinds of fungi are abundant in decaying plant material, with their microscopic extensions permeating the leaf litter, soil humus, and rotting logs. The capacity to break down cellulose is an ancient biochemical pathway and certainly assisted fungi as they first colonized the land in concert with advanced plants. However, the tendency for certain groups of fungi to engage in sustained alcoholic fermentation turned up only much later in evolutionary time and is found in only a small subset of all yeast species. Not surprisingly, the most widespread of these today is the one co-opted by humans for brewing and winemaking, *Saccharomyces cerevisiae.* This species, which is also used in bread-making, has essentially been domesticated through its thousands of years of association with cuisine. In tropical environments, many other species of fermenting yeasts can also be found in association with ripe and over-ripe fruit. The common feature of all such yeast assemblages is competition with bacteria and the ensuing production of alcohol.

But what exactly is the chemical process that yields such an interesting molecule? After considerable speculation in the mid-nineteenth century, it was Louis Pasteur who first proved experimentally that fermentation requires both the presence of sugars and the metabolic activity of yeast as a necessary biological participant. Intriguingly, yeasts can produce alcohol in the complete absence of oxygen. Such fermentation is accordingly known as an anaerobic process, and was identified as such by Pasteur when he termed it to be "la vie sans l'air." This was a remarkable finding that, in retrospect, has important consequences for our understanding of the evolutionary origins of this metabolic pathway. Fermentation to yield a variety of non-alcoholic compounds is actually an ancient biochemical process used by many different kinds of microbes to produce energy-rich compounds. Plants can also engage in anaerobic fermentation under certain conditions, as when roots become submerged. On the geological timescale, this metabolic pathway well preceded the origins of sugar-rich fruits. However, the flowering plants of the Cretaceous provided within their fruits a new arena of simple carbohydrates well suited for fermentation and the associated generation of alcohol. When sugar levels are very low, fermenting yeasts produce no alcohol. Sugars are simply burned up aerobically to contribute to growth and metabolism. Sugar at concentrations greater than 0.1%, however, suppresses such activity via a well-studied biochemical switching mechanism that turns on the pathway of alcohol production. Increasing sugar concentrations thus elicit anaerobic fermentation even when oxygen is present. For yeast cells growing within watery fruit pulp, conditions are probably oxygen-deprived in any event, and fermentation is the order of the day. Alcohol builds up accordingly.

In fact, the fermentation of fruit sugars by yeasts yields a number of different alcohols and end products, including glycerol, acetic acid (i.e., vinegar), lactic acid, and numerous aromatic compounds. It is the short-chained ethanol molecule that is the predominant alcohol, however, contributing about 90% of the total yield. Additional compounds, including the fusel oils (so-called higher alcohols with longer chain molecular structures), contribute to the flavor and bouquet of alcoholic beverages, but these are clearly second-order participants. If we are to look for a good explanation for anaerobic fermentation, then we should concentrate on alcohol per se. However, production of this molecule by yeast is a surprising result given that complete oxidation of a sugar molecule (in this case, glucose) yields as many as thirty-eight molecules of energy-rich adenosine triphosphate, whereas fermentation of glucose to alcohol (i.e., ethanol) yields a paltry two molecules. Alcohol molecules thus retain high energetic content, an outcome perhaps most obvious in the interesting if perhaps grotesque phenomenon of the beer gut. Drinking a lot of booze clearly packs on the intrinsic calories of the alcohol molecule.

Surprisingly, only about 5% of potential metabolic yield is realized by yeast using anaerobic fermentation, relative to what they could achieve with full oxidation of the sugar molecules. Why then do yeast cells bother to produce alcohol at all? They could presumably get much more energy by fully oxidizing all available carbohydrate, but evolution has apparently preferred an alternative solution that, in a broader perspective, must yield greater long-term results. Here is where a historical perspective on biology is critical for explaining seemingly inefficient, energetically disadvantageous, or simply foolish behaviors. Using DNA sequencing and sophisticated methods of evolutionary analysis, it is possible to reconstruct the history of the fungi. This exercise places the origin of the fermenting yeasts back to the mid-Cretaceous, roughly corresponding to the period about 120 million years ago when flowering plants first arose and began to produce fleshy and sugar-rich fruits. Although there is some uncertainty as to the correct rates of the molecular clocks used in such estimates of deep time, the broad temporal congruence of these two events is suggestive. There must be some link between yeast fermentation and the internal environment of the fruits within which they thrive.

What then might be the non-energetic benefits accruing to yeasts that produce alcohol? To date, the best explanation is that this alcohol acts to inhibit microbial competitors. Initially, sugar concentrations within ripe fruits are high but yeast populations are low. Growing yeasts typically produce much more alcohol than those at rest, and both yeast population densities and alcohol levels rapidly climb. Competing bacteria, in spite of their much faster growth rates relative to those of yeast, are at a disadvantage because their capacity to reproduce is progressively inhibited as alcohol concentration increases. Substantial osmotic stress associated with both alcohol and high sugar concentrations also dehydrates bacterial cells. By contrast, yeasts have a much greater tolerance for alcohol. In fact, yeast growth is inhibited only at levels of 10 to 14% (i.e., at levels typical of many wines), depending on the kind of yeast and the particular conditions of temperature and pH that surround it. Bacteria, by contrast, are killed off by much lower alcohol levels, mostly because of their much simpler cell membranes.

Yeasts can thus beat up on the bacterial competition through a form of chemical warfare.

Moreover, the interior of unripe fruits also tends to be somewhat acidic to deter premature consumption by birds, mammals, and possibly some microbes as well. Here, the yeasts have a real advantage. Growth rates of bacteria are substantially inhibited at pH levels below 6.0, whereas fermenting yeasts do better at the much lower values characteristic of unripe fruit (typically pH values of 2 to 5). Therefore, yeast can gain a larger share of the carbohydrate resources of a fruit by suppressing bacterial growth. This relative advantage for yeast under these acidic conditions, when coupled with the inhibitory effects of alcohol on bacteria, yields an overall competitive asymmetry for yeast populations. Alcohol levels correspondingly rise, sugars decline from the levels characterizing peak ripeness as they are consumed by yeasts, and the bacteria are held in check. Secondarily, the yeasts will then shift to burning up the alcohol molecules themselves when sugars are no longer present. The fruit itself, assuming it has not been consumed by an animal sometime during this process, will then ultimately succumb to bacterial rot. Fermenting fungi thus act as the primary agents of decay, assuming that they have been able to colonize the fruit to begin with.

Not surprisingly, the life cycle of yeasts predisposes them to development within and on fruit. In the natural world, fermenting yeasts will invade a sugary habitat and grow rapidly via asexual budding. Intermittent sexual reproduction also produces large numbers of spores. These spores are small, on the order of several microns, and once released from a substrate, are readily wafted about in the air. They thus can easily inoculate developing fruits from the surface and work their way inwards. Fruit flies and other insects such as bees and wasps also inadvertently move yeasts from plant to plant, which may select for volatile attractants such as alcohol, which promotes such visits. More cleverly, many fungal spores land first on flowers and are then encapsulated into the fruit tissue as it develops from the fertilized flower. This latent infection yields, for example, avocados rotted from the inside out, an occasionally disconcerting experience for those who purchase otherwise unblemished fruits from the grocery. Many kinds of yeasts can typically be found either within or upon the surface of fruits in both the temperate and the tropical zones. Bacteria omnipresent in the air and on plant surfaces are similarly opportunistic colonists. Yeasts and bacteria may thus grow side by side within fruit, although growth rates of both are low while the fruit remains unripe.

Plants are not oblivious, in evolutionary terms, to the presence of these microbial agents of destruction. Many fruits are fragrant and aromatic precisely because of the presence of many antimicrobial compounds. Consider, for example, the sweet perfume of a ripe mango, the refreshing zest of lemons and oranges, and the many other fragrant aromas of tropical fruits. Because of the intrinsic reproductive value of developing seeds, there has been considerable evolutionary pressure on plants to express antimicrobial compounds within their fruit. These chemical defenses must coexist with the palatable and nutrient-rich pulp which animals will consume. There can also be intense competition for sugars among different yeasts within the same fruit. So-called killer strains sometimes turn up and secrete toxic molecules which serve to neutralize other yeast strains growing on the same substrate. These killers can occasionally colonize and spoil grape batches collected for wine fermentations and are thus feared by vintners. Similarly, some kinds of bacteria can colonize alcoholic fermentations and convert them to vinegar. This microbial finale is often the endpoint of an opened but subsequently neglected bottle of wine. Refrigeration dramatically slows this process and, along with stoppering to impede oxidation, is recommended for partially consumed bottles.

The role of yeasts in influencing the likelihood that a fruit will be sensed at a distance and ultimately consumed by an animal is difficult to assess. Overall, sustained microbial rot tends to reduce fruit attractiveness, but the parallel time course of alcohol production as a fruit ripens and then decays may offset this trend. Yeasts are more likely to be producing alcohol early on in a ripening sequence, given that simple sugars become progressively more available. And the numerical dominance of yeast relative to bacterial populations likely sustains the palatability of a fruit, as it does with rising bread. Also, if alcohol emanating from a fruit serves to attract vertebrate dispersers, it may be in the plant's best interests to tolerate a certain amount of yeast growth, as the small loss in carbohydrates may be more than offset by an increased chance of discovery, consumption, and subsequent dispersal. Increased alcohol levels also increase ecological shelf life, as it were, through inhibition of bacterial growth and the postponement of associated rotting. Prior to actually tasting a ripe and fermenting fruit, it can be difficult to assess the quality of the contents. Animals (including us) will also sometimes eat only part of a ripe fruit and reject the rest. In evolutionary terms, plants will try to minimize their investment in sugars and other enticements to consume, whereas animals try to increase their net caloric gain per unit of time when foraging. Evolution of these kinds of mutualistic interactions often acts in reciprocal fashion and ultimately can yield more and more complex strategies of both entrapment and evasion (unfolding over millions of years) by all participants.

Another possibility is that sufficiently high alcohol levels may actually deter consumption if certain animals find the taste aversive. Such an outcome might facilitate the growth of yeasts and ultimately their reproduction, but would work against the genetic interests of the plant if the seeds are never dispersed. In any event, there is no evidence to date demonstrating negative effects of naturally occurring alcohol on fruit consumption. The relative importance of alcohol in either attracting or deterring animal consumers is likely to be dependent on concentration and how it varies through time. The particular taxonomic identities of the plant, animal, and fungal participants in this three-way interaction will also be reflected in variable patterns of fruit ripening and the corresponding expression of alcohol. Given the huge number of flowering plants that produce sugar-rich fruit, together with the myriad kinds of mammals and birds that consume them, a wide diversity of behavioral and ecological outcomes can be expected. The sensory physiologies of animal frugivores (i.e., their capacity to smell and taste numerous fruit-specific flavors, including alcohol) should be similarly variable.

When humans use yeast to produce beer and wine, we are typically limited to an alcohol content below 15% because of alcohol's inhibitory effects on the yeast's biochemical functions, and ultimately on its growth. Higher concentrations of alcohol can only be obtained via the chemical process of distillation, as discussed in detail in chapter 5. Moderate concentrations of alcohol are nonetheless highly effective as food preservatives, given the protective effect of the molecule against bacterial spoilage. Humans worldwide have converged on the partial fermentation of vegetables (e.g., sauerkraut) and milk products (e.g., cheese) for long-term storage. We even use ethanol or related compounds as a clinical swab prior to injections to kill off bacteria on our skin, indirectly enlisting the by-products of fungal metabolism to aid in our own microbial wars.

Much of what we know of yeasts and alcohol production originated in the cultural processes of brewing and wine-making, as developed over millennia. Now, by using yeast to ferment mostly corn, wheat, and sugar cane, humans produce about thirty billion liters of alcohol every year. This amount represents the second-largest biotechnological product in the world, exceeded only by the harvest of both farmed and natural timber. About a quarter of our efforts in fermentation goes towards alcoholic beverages, and the remainder is used as industrial alcohol and motor fuel. Current interest in alternative fuels and gasoline additives will only increase this latter usage, and the microbial production of alcohol from plant products has a potentially lucrative future worldwide. My own university, by way of example, recently accepted hundreds of millions of dollars from British Petroleum to establish an institute for research into the commercial production of such biofuels. This impressive attempt at the greenwashing of oil profits amply brought home the energetic impact of the ethanol molecule. But what of alcohol production in the natural world? How often do yeasts and alcohol actually build up to significant levels within fruit? And are there other biological circumstances whereby alcoholic fermentation might occur in sugar-rich pulp or other kinds of biological products (e.g., nectar)? What do we know about the natural ecology of fermenting yeast on fruits in the real world, and particularly in the tropics?

A MOST DELICIOUS LIQUOR

Yeasts grow best in sugar-rich environments, and their growth is facilitated by higher temperatures and relative humidities. An obvious inference, therefore, is that ripe fruits in wet tropical environments will contain the most alcohol relative to fruits in other regions of the world. To date, however, most available information on alcohol content pertains only to domesticated fruits found within the temperate zone, and particularly to grapes. These studies indicate concentrations in decomposing fruit that range widely, from trace quantities to values as high as 5%. Some of these studies must have been great fun, as they involved sampling grapes and wine seeps from the discard piles of vineyards. Although these are obviously artificial contexts for alcohol production, the local fruit flies have clearly become accustomed to its near unlimited availability, as we shall see in the next chapter. Some limited data on alcohol content are also available for bananas, strawberries, and some other kinds of commercial fruit crops.

But such situations tell us very little about the natural ecology of fermentation and alcohol production by yeasts. Humans have domesticated grapes and other fruits for millennia. Through the process of artificial selection, we have dramatically altered their texture, taste, chemical composition, and appearance relative to their natural precursors. Because we don't like our fruits to be spoiled in any way, no matter how superficially, we apply lots of pesticides and fungicides during the ripening period. Artificial selection may also result in greater chemical defenses against microbial pathogens like yeast and bacteria, either for the sake of visual appearances, or simply to increase overall yield. So the data on alcohol levels within grapes and other agricultural crops, although suggestive, tell us basically nothing about what might be happening in the natural world with wild fruit and fermentation. The yeast strains that infect domesticated crops, as well as those used in breweries, also differ substantially in growth characteristics and in their physiological ability to tolerate alcohol relative to wild yeasts.

If we instead consider ripening profiles and associated alcohol levels within naturally occurring fruits, the data are remarkably sparse. When I first started working on this problem, I actually could find no information whatsoever on the alcohol content of fruits eaten by wild primates. To offset this absence, I made a number of relevant measurements on the relatively large fruits of the palm *Astrocaryum standleyanum* (see plate 2). Fruits of varying ripeness conditions were obtained either from palm trees or from the forest floor on Barro Colorado Island in Panama and were then analyzed at the on-site field station run by the Smithsonian Institution. Categorization of ripeness (i.e., unripe, ripe, and over-ripe) was made a priori and somewhat arbitrarily conformed to my own personal preference, although surface color was clearly different among the three groups. The unripe green fruits contained no measurable alcohol, but the situation was very different for ripe and over-ripe fruits. Alcohol content for pulp of the former averaged 0.6%, whereas that for the latter was a relatively high 4.5%. Both ripe and over-ripe fruits also exhibited, in aggregate, a wide range of concentrations, indicating that color per se does not necessarily indicate the extent of fungal infestation.

Nonetheless, alcohol levels within these fruits are non-trivial and correspond roughly to those found in weak beer. Given that 40% or so of any given palm fruit is pulp (as distinct from seed and husk), this represents a potentially substantial exposure to alcohol for any animal that consumes large amounts. Mammalian frugivores can eat up to 5 to 10% of their body weight per day in fruit, so daily low-level exposure to dietary alcohol should be expected in many species. I also regularly see large, iridescently blue *Morpho* butterflies sucking up goopy fermented liquid from these fallen palm fruits. At first glance, the measured concentrations of alcohol in the pulp may seem to be insufficient to enable actual intoxication. But as we all know, it's how much one drinks as well as the concentration that determines the physiological hit of alcohol and ultimately drunkenness. Here is where the data are really limited—we know essentially nothing about typical feeding rates for the animals that are known to eat these palm fruits. Still, this set of measurements was a first pass at the interesting question of alcohol levels in the wild, and it turned up fairly encouraging results.

Additional measurements on fruit alcohol content have been made by my colleagues Carmi Korine, Berry Pinshow, and Francisco Sánchez at Ben-Gurion University, working on the flora of the Negev desert in southern Israel. It is perhaps surprising to hear about fruit growing at all in such dry conditions, but many local plants (such as mistletoes) seasonally provide large numbers of smallish fruit to migratory birds. Fruit bats also visit figs and palms throughout the Mediterranean and other semiarid regions of the world. For four species of ripe fruit in the Negev, alcohol content of the pulp was about 0.44%, somewhat lower than that for the Panamanian palm fruits but still potentially substantial depending on the amount consumed. Moreover, Nate Dominy (then at the University of California, Santa Cruz) measured alcohol content for seven tropical fruit species in Singapore and found values ranging from 0.12 to 0.42%. Clearly, fruits in the wild contain non-negligible amounts of alcohol, although the exact concentration will depend on the locale, season, ripeness, and of course taxonomic identity of the fruit in question. A key issue now is to determine the extent to which fermentation and alcohol levels are correlated with ripening profiles and with the relative attractiveness to fruit-eating birds and mammals. Ultimately, we would like to know how much alcohol is consumed in a feeding bout by these animals, and over what time period, so as to be able to estimate directly their physiological exposure.

For the ripe and over-ripe palm fruits in Panama, alcohol levels and sugar content were inversely correlated, exactly as would be expected given that yeasts progressively ferment and deplete all available sugars. What we have no information on, however, is the simultaneous interplay between bacterial and yeast populations as a fruit ripens. The fate of alcohol produced within the fruit should strongly influence relative growth of these two competing groups. If alcohol indeed functions as an antibacterial killing agent, then yeast populations should increase up to maximum values well before the bacteria catch up, and then only slowly decline. For small fruits with large surface-area-to-volume ratios, however, alcohol will diffuse relatively quickly out of the fruit's interior and into the surrounding atmosphere, yielding lower internal concentrations. Bacteria will probably do better under such circumstances. Larger fruits, by contrast, will retain more of the alcohol produced by yeasts, and thus would be predicted to have a higher alcohol content under comparable conditions. We might expect to see a longer delay in bacterial buildup in large fruits, which, among other effects, will tend to prolong their attractiveness and availability to animal frugivores.

Other biological contexts can be also identified for sugar-rich solutions that ferment in the wild. Flowers typically (but not always) secrete nectar to attract pollinators and are often found in warm sunlit environments conducive to microbial growth. A recent study in Malaysia of a large palm flower found the nectar to be colonized by fermenting yeast, along with alcohol levels up to levels of 3.8% (and an average content of 0.6%). The interesting effects of this nectar on the various nocturnal mammals that consume it are discussed in chapter 3. The potential role of alcohol in attracting animal pollinators to flowers more generally has never been examined, although honeybees and wasps would be interesting subjects for such investigations. Many different birds (e.g., hummingbirds in the New World and sunbirds in the Old World) also feed primarily on floral nectar, as do the New World flower bats and many tree-dwelling mammals worldwide. Some nectars, moreover, contain low concentrations of chemical compounds that inhibit microbial activity, suggesting that fermentation may degrade sugar rewards and thus pose a real problem for flowers. Alternatively, alcohol odors might preferentially attract certain kinds of pollinators, and the evolutionary interactions between flowers and yeasts might accordingly be more complex. For example, bumblebees have recently been shown to prefer flowers containing nectar that has been artificially inoculated with yeast. The number of seeds in the flower are reduced, but yeast dispersal by the insects may be enhanced. Floral warming through yeast metabolism could also be attractive to some pollinators. Remarkably, neither attractive nor deterrent effects of alcohol alone on flower visits and pollinator preferences have been evaluated under natural conditions.

Some plants with flowers also bear small cup-like structures called extrafloral nectaries on their stems and leaf bases that secrete low-concentration sugar solution in order to attract ants (see plate 4). Ants feed on these sugars for energetic reward and in turn defend the plant against herbivores such as caterpillars. Although never studied in this regard, these nectaries may also contain some low-concentration alcohol from time to time. Similarly, honeydew is exuded by aphids and other related species to attract ants that serve as bodyguards to these insects. As with nectar, honeydew contains low-concentration sugars and may similarly ferment prior to its consumption by the host-tending ants. Because microbial growth speeds up dramatically at higher temperatures, alcohol can build up to physiologically relevant concentrations in just minutes if yeast cells are present and if the underlying biochemical conditions are appropriate. Ants boozing on honeydew have yet to be identified but clearly can't be ruled out given what we know about the natural history and ecology of this system.

Overall, it is clear that warm tropical environments are the likeliest places for naturally occurring fermentations. Yeast spores are borne by the winds and can land pretty much anywhere. And the plant-derived sugars they feed on are abundant within fruit, nectar, and even extrafloral nectaries. Fermentation and alcohol buildup are thus inevitable. For many animals, this is a great outcome given that they can potentially sense alcohol vapor over long distances and can reliably associate the odor of booze with substantial energetic reward. Fruits in the tropics are of particular interest in this regard given their wide taxonomic diversity, along with the numerous and physically large bird and mammal species that eat these fruits and literally carry out (in their gut) the services of seed dispersal. Real data on fruit-alcohol content are few and far between, however. Many animals in the tropical rainforest are likely to be consuming alcohol, but both the timing and extent of such exposure are unclear. Nonetheless, booze is clearly out there in the real world, and the next logical step is to look at the behavioral responses of different kinds of animals to alcohol. As we shall see, both fruit flies and barflies (the human kind) are well-studied in this regard, but the natural biology of alcohol exposure in all other animals is otherwise wide open and ripe for exploratory investigations.

CHAPTER THREE

On the Inebriation of Elephants

We are all familiar with the human drunk, and with the full range of her or his behaviors. These can range from the merely entertaining to embarrassing, damaging, and even death eliciting. But are there comparable outcomes in the animal kingdom? An Associated Press story published online in 2002, for example, relates elephants marching through villages in Assam in search of illicit stills, which they broke open in order to quaff home brew. They then ran amok in a drunken rampage, even killing villagers. Similarly, numerous accounts of inebriated mammals and birds relate the consumption of either fermenting foodstuffs (such as bread dough) or alcohol-laden fruit, followed by apparently drunken comportment. This anecdotal and often humorous literature is, however, very difficult to interpret scientifically.

Are there any real data demonstrating alcohol intoxication in the wild? And are there any evolutionary expectations for animal physiology and behavior if low-level exposure to booze is an inevitable consequence of a fruit-based diet?

THE ANIMAL INEBRIATE

By many anecdotal accounts, drunkenness would seem to occur frequently in the animal kingdom. Cavorting groups of inebriated baboons, sozzled chimps falling out of trees, and birds too drunk to fly have all been described by naturalists, interested bystanders, and a more voyeuristically oriented popular media. Many stories hint at behavioral similarity between the drunken beast and inebriated humans. And sometimes the observers themselves might have been drinking, based on the tone of the reports. Most suggestive is a widely distributed sequence from the film *Animals Are Beautiful People* (1974) which portrays staggering, cavorting, and apparently natural drunkenness in South African mammals ranging from baboons to elephants and zebra. It was subsequently revealed that these animals had been either fed excessively high levels of liquid alcohol or had been injected with a veterinary anesthetic in order to elicit such behaviors, which otherwise have never been observed in the wild.

Nonetheless, some stories of alcoholic animals contain more than a hint of scientific truth. In 1990, veterinarians measured alcohol levels in two recently deceased cedar waxwings that had eaten hawthorn fruits and then tragically fallen from a rooftop. Alcohol concentrations in the livers and crops of these birds were ten to one hundred times higher than those measured in control bird species, suggesting a high level of alcohol ingestion. Cedar waxwings seem to be at particularly high risk in this regard given their repeated appearance in the popular literature; multiple reports in North America have them flying drunkenly into windows and buildings. Fruit-eating birds in the temperate zone may also be particularly susceptible to inebriation when they consume berries fermenting in the spring thaw. A 2012 report from Cumbria in the United Kingdom similarly reported high levels of alcohol in dead blackbirds and redwings, consistent with lethal intoxication.

By contrast, tales of drunken animals in the tropics tend to focus on much larger species such as elephants, warthogs, and giraffes. In southern Africa, these animals are often reported to consume large quantities of fruit from a common and widespread tree called the marula. The yellow fruits of the marula tree are about four centimeters long; when ripe, large numbers of them fall to the ground, where they subsequently ferment and are consumed by the local fauna. Local peoples also use marula fruit for food and for moonshine production of fermented beverages. Elephants figure prominently in stories of marula-induced drunkenness, and the commercially produced South African Amarula Liqueur even features a proud pachyderm on the bottle's label. Marula fruit were also ostensibly the source of natural alcohol for the animals portrayed in the film *Animals Are Beautiful People*.

But how many marula fruits must an elephant actually consume to be inebriated? In 2006, a team of physiologists estimated that the number would have to be excessively high. Given reasonable assumptions about the alcohol content of the fruits, likely ingestion rates, and degradation of alcohol within the elephant's digestive system, this group concluded that the rates of natural dietary exposure to alcohol would be at least a factor of four below that required to effect overt intoxication. Even ingestion of a substantial thirty kilograms of marula fruit, corresponding to a daily meal equal to about 1% of an elephant's body mass, would be inadequate. And this estimate makes the conservative assumption that the meal was not diluted by any water drunk over the same time period. Elephants are obviously very large, and unrealistically high numbers of fruit would be necessary to attain meaningful alcohol load. Under natural conditions, drunkenness in elephants, and presumably in other large mammals, would be very unlikely.

Nonetheless, biologists may well have missed the broader significance of even rare outbursts of drunken behavior, particularly given the diversity of fruit- and nectar-consuming animals that are exposed to alcohol. In the humid tropics, for example, many butterflies feed on fallen and decomposing fruits, rather than on nectar from flowers (see plate 5). Lepidopterists have traditionally used fruits and other fermenting substances to attract both butterflies and moths, and at least one published account suggests natural inebriation of a butterfly feeding on rotting fruit. Australian lorikeets have similarly been reported to become drunk while feeding on fermented nectar, ultimately becoming unable to fly. In fact, it might be particularly dangerous to drink alcohol while flying. A series of studies on fruit bats in the Negev desert suggested that although they can sense alcohol in solution at very low levels, aqueous concentrations above 1% are actually avoided. Nocturnal flights to and from communal roosts are obligatory for these large bats, and any inability to fly would involve substantial risks for them, and indeed for any flying animal that has but limited mobility on the ground. Behavioral responses to alcohol are thus likely to vary with the animal species in question and with various aspects of its physiology and natural ecology.

Further confounding the interpretation of apparently drunken animals, such individuals may in fact be intoxicated through the pharmacological action of plant alkaloids or other secondary compounds that can be concentrated within fruits or leaves. Think, for example, of the aromatic flavors of many tropical fruits, or of the acerbic and rough taste of unripe stone fruit (e.g., green peaches). The associated chemical compounds are widespread in fruits and may yield overt behavioral effects at low physiological concentrations. Similarly, nicotine is found in the nectar of tobacco flowers and can intoxicate insects that feed on these plants. Because we have no data on blood-alcohol concentrations for apparently inebriated frugivores and nectar feeders, it has been impossible thus far to differentiate drunkenness from other possible toxicological effects. Nonetheless, if animal inebriation were widespread, then it is clear that this phenomenon would have been well documented and analyzed over the last several hundred years of biological science. Given the wide taxonomic range of birds and mammals potentially exposed to alcohol via ingestion of fruit and nectar, what is instead surprising is the relative scarcity of such reports.

Also possible, of course, is that animals do indeed consume lots of alcohol in the wild but then degrade it quickly using effective enzymatic machinery. A remarkable study published in 2008 nicely illustrates this effect for nectar-derived alcohol in the Malaysian rainforest. The large flowers of the bertram palm produce copious amounts of frothing and fermenting nectar, upon which a variety of nocturnal mammals routinely feed. The researchers identified the yeast involved in fermentation, measured the alcohol content of the nectar, and quantified the amounts consumed by individuals. Application of physiological models to these data suggested that for one species of mammal closely related to primates (the pen-tailed treeshrew), blood-alcohol levels would exceed physiological levels associated with human drunkenness about one day out of three. Yet drunken behavior was never observed in any of the animals feeding on the nectar. Most importantly, a secondary metabolite of alcohol degradation (called ethyl glucuronide) was identified in three species of mammals that routinely drank the fermented nectar, but not in some of the other local mammals that feed mostly on non-fermenting food (e.g., long-tailed macaques). Ethyl glucuronide otherwise turns up at non-trivial concentrations only in human alcoholics. The nectar-feeders are thus consuming large amounts of alcohol but are either degrading it quickly or are much less susceptible to its inebriating effects.

In sum, behavioral observations to date suggest only occasional alcohol intoxication in animals. The phenomenon is rare, and it is not difficult to envision that selection would act against individuals who routinely become drunk. Simply put, other daily activities would suffer in a world that takes natural selection seriously. Animals in the real world routinely face hunger, diseases, and predators. Any behavior that compromises performance will be selected against. We should accordingly not expect drunkenness to be an everyday event, although gorging on fermented fruits during periods of food limitation might occasionally be expected. Similarly, most people today who occasionally drink alcohol do not become dangerously intoxicated. However, a more pervasive role of alcohol may lie in the fairly low-level concentrations that animals experience through routine consumption of, or even physical residence within, fermenting fruit pulp. And although it may seem unlikely, such is the habitat of larvae of the appropriately named fruit fly (i.e., flies in the family Drosophilidae, the most famous of which is the genus *Drosophila*). Studies of these insects and of the role of alcohol in their life cycle have yielded great insight into both how the molecule is biochemically degraded and the long-term evolutionary consequences of natural exposure.

FLIES LIKE IT TOO

Fruits placed outdoors, in either tropical or temperate climates, exude odors that attract insects. As a ripe fruit progressively decays and begins to rot, the odors of alcohol and other fermentation products will bring in numerous fruit flies. By flying upwind once they encounter an appropriate smell, these flies rapidly home in on a suitable target upon which to lay eggs.

This response, along with the neurophysiological capacity of fruit flies to sense alcohol molecules, has been well characterized in laboratory experiments. In a wet tropical forest, ripe fruit is rapidly covered by an assemblage of fruit flies jockeying for position (see plate 6). Female flies first choose among and mate with males and then lay their eggs in the pulp. For these flies, the scent of alcohol reliably indicates the presence of a sugar-rich and yeast-laden medium that will nutritionally sustain their developing larvae. That evolutionary selection has been strong on this behavioral response is well indicated by the use of fermenting odors alone by some flowers to attract pollinators. Without providing a nectar reward, but through copious use of volatile odors mimicking fermentation, the Solomon's lily and other Mediterranean flower species attract fruit flies to effect pollination but are spared having to provide sugars as compensation. Flies simply like to hang out around these odors of booze. Alcohol molecules, in other words, are a powerful attractive force in nature.

Adult fruit flies, in addition to mating on ripe fruits or other similar substrates (e.g., rotting cactus tissue), can also feed directly on the fermenting carbohydrates. In laboratory experiments, *Drosophila* flies also prefer to consume pulp that has been supplemented with alcohol. Direct neurophysiological measurements on their chemical receptors also demonstrate that these flies can both smell alcohol molecules in the air and taste them with sensory structures on their feet and proboscis. Interestingly, their preference for booze persists even when a distasteful compound is added to alcohol-supplemented food. Flies, moreover, exhibit an increasing preference for alcohol-containing food over time, and also tend to consume greater amounts given repeated exposure. Relapse to high levels of alcohol consumption can also occur following investigator-enforced abstinence. And most remarkably, male flies deprived of mating opportunities exhibit an increased preference for alcohol. In aggregate, the parallels between these behaviors and modern human consumption of alcohol are substantial. Patterns of increased drinking through time, as well as an important role for alcohol in social contexts, should be familiar to all of us. But are these various aspects of fly addictive responses physiologically identical to those in modern humans? The structure and composition of the nervous system in insects and mammals differ in many important ways, for example, so it is important to not anthropomorphically project too much into these kinds of comparative results. Nonetheless, fruit flies represent an important experimental system with which to evaluate natural responses to alcohol and to investigate potential genetic underpinnings to behavioral preferences.

Once fly eggs hatch within a ripe fruit, the larvae then cruise through the decaying pulp, eating the associated fruit sugars and yeast spores and also metabolizing the alcohol produced by the adult yeasts. Exposure to alcohol in this mushy domain is thus considerable and influences both the behavior of adult fruit flies and the physiology of the larvae. Given such exposure to alcohol through their entire life cycle, it is not surprising that fruit flies express in abundance the enzymes necessary to metabolize it. Alcohol is first acted on by the alcohol dehydrogenase enzyme (ADH), which yields an intermediate

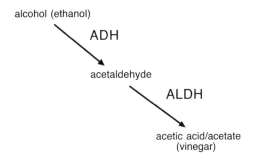

Figure 1. Biochemical action of the ADH (alcohol dehydrogenase) and ALDH (acetaldehyde dehydrogenase) enzymes on the alcohol molecule.

product termed acetaldehyde (see figure 1). Although toxic and even carcinogenic (to humans) at sufficiently high concentrations, acetaldehyde is quickly degraded by the enzyme aldehyde dehydrogenase (ALDH). The end product is the acetate molecule, which then enters a well-studied biochemical pathway (glycolysis) to produce several kinds of energy-rich molecules. These, in turn, are oxidized to yield even more energy at what is the endpoint of this complicated process of alcohol metabolism.

Starting in the 1960s, the amino acid sequences for the ADH and ALDH enzymes in wild fruit fly populations were systematically analyzed to characterize underlying patterns of genetic variation. The results were surprising in that many variants were found to exist in different regions of the world, and in that those individuals with faster acting enzymes exhibited greater adult tolerance for exposure to alcohol. Some of these studies were conducted in temperate-zone vineyards and at wine seepages from fermentation barrels where alcohol levels were artificially high. The flies taking advantage of these resources had, apparently within decades, evolved much better enzymes to enable higher levels of alcohol tolerance in both the larvae and the adults. Similarly, fruit flies living in warmer latitudes along the eastern coast of Australia exhibit faster enzymatic metabolism of alcohol, presumably because fermentation and its fruit-derived alcohols are more prevalent at higher air temperatures. These responses can also be elicited in the lab using flies evolved over many generations on alcohol-supplemented media. These and other responses that have evolved over millions of years can be deciphered by directly analyzing the different genes that encode for ADH and ALDH, yielding DNA sequences for comparison among different populations and different species of fruit flies. This approach continues to be fruitful territory for the field of evolutionary genetics and well demonstrates the wild and powerful forces of natural selection in molding the genome.

Although alcohol levels within fruit are fairly low, and flies usually do not become inebriated in the wild, adult fruit flies do get drunk when exposed to sufficiently high levels of alcohol in the lab. Not surprisingly, they wobble about and fall down in impressive exhibits of reduced motor control, just as human drunks stagger and collapse. For fruit flies, this behavioral response has been used to great advantage in a wonderful experimental apparatus termed the inebriometer. Flies are placed above a liquid pool of high-concentration alcohol and voluntarily fly upwards towards a light source until they land and then pass out on one of a series of stacked funnels. Higher fliers are those individuals less susceptible to intoxication, and these can be collected and then analyzed for genetic composition. Hundreds of individuals can be tested in this way at once, permitting rapid screening for genetic variants which possess a particularly high resistance to alcohol. By then sequencing the associated genes found in these mutant individuals, the underlying basis of tolerance to alcohol can be identified and placed in the broader context of fly genomics and biology. This approach enables those genes that are at play in molding behavioral and metabolic responses to alcohol to be studied among different species and among different groups of fruit flies with naturally variable levels of exposure.

These studies have revealed, quite remarkably, that the molecular mechanisms of inebriation in fruit flies are similar to those found in mammals. For example, certain cell signaling pathways, as well as assemblages of neurons that utilize the neurotransmitter dopamine, have been implicated in the alcohol responses of both fruit flies and rodents. Specific genes associated with susceptibility have also been identified. The intriguingly named *cheapdate* mutant is, for example, particularly sensitive to the inebriating effects of alcohol, whereas the *happyhour* mutant exhibits increased resistance. By contrast, short-term tolerance to alcohol (i.e., a reduced sensitivity) is correlated with the mutation termed *hangover*. And in a remarkable parallel to human behavior, male flies whose courtship advances have been rejected by females tend to prefer alcohol-enhanced food, and this effect is mediated by a small protein involved in neural reward circuitry.

Importantly, such molecular and genetic influences on fruit fly physiology act independently of the capacity to metabolize alcohol via the ADH and ALDH enzymes. By pinpointing such genetic correlates to the susceptibility for inebriation and alcohol tolerance in fruit flies, equivalent pathways can potentially be identified in humans and then used as targets for pharmacological intervention. Cell signaling pathways that indicate reward in response to alcohol exposure, for example, could be selectively disrupted by specific classes of drugs that might also work for humans. But the responses of fruit flies to alcohol and their molecular underpinnings go far beyond the immediate effects of drunkenness. Somewhat paradoxically, some long-term effects of alcohol exposure, in addition to the deleterious outcome of addiction, may actually be positive in a variety of ways for both animal and human health.

A POISON THAT HEALS?

Although high concentrations of alcohol are clearly toxic, routine exposure to much lower concentrations may have a very different outcome. Substantial evidence indicates that limited exposure to alcohol can actually be beneficial in a variety of ways. A remarkable study published in 1926 first documented this effect experimentally, albeit for the improbable choice of the domestic chicken. Atmospheric alcohol vapors both reduced mortality and increased overall life span for young and old chickens alike. Although this approach was never implemented in the poultry industry, it nicely illustrates a more general point. Many chemical compounds that are toxic at high concentrations can be beneficial at low dosages. And exposure can derive either from an environmental source (as in the case of the chickens and alcohol vapor) or via direct ingestion of particular chemicals. Vitamins and essential minerals, for example, are compounds essential for life that can nonetheless be dangerous when overconsumed.

This outcome is an example of an important concept in toxicology termed hormesis, whereby low-level exposure to naturally occurring substances is often beneficial. By contrast, negative effects will ensue if there is either no exposure or exposure to abnormally high and damaging levels. Hormesis often reflects evolutionary adaptation to compounds which occur in the environment at low concentrations and which can be co-opted for physiological benefit. In abnormal situations of high-level exposure, however, these chemicals are poisonous. Animals have never historically consumed high concentrations of alcohol given the values typically found in fermenting fruit (chapter 2), but their regular dietary exposure may otherwise afford real benefits. In turn, natural selection will act to favor the evolution of genetically based behaviors and physiological capacities that take advantage of the alcohol molecule.

These effects are pronounced in the case of fruit flies. Adult flies exposed to vapor concentrations of up to 4% exhibit an increased life span relative to those exposed to higher concentrations or to water vapor alone (see figure 2). Acetaldehyde, the intermediate product of alcohol metabolism, is similarly beneficial at very low concentrations, both for fruit fly larvae and also for some species of much smaller wasp parasites that can develop inside them. Other kinds of internal wasp parasites, by contrast, are killed off by higher alcohol concentrations that fruit fly larvae intentionally seek out when infected. This behavior suggests self-medication by the larvae as they burrow through fermenting pulp, and it indicates the multiple roles of the alcohol molecule in fruit fly biology. Moreover, adult female flies prefer to lay their eggs in ethanol-rich medium if they visually sense the presence of wasp para-

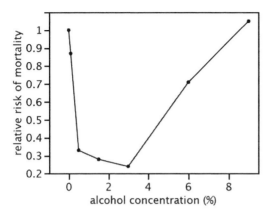

Figure 2. Relative risk of mortality for the fruit fly *Drosophila melanogaster* as a function of exposure to alcohol vapor of different concentrations (data from Peter Parsons, "Ecobehavioral genetics: habitats and colonists," *Annual Review of Ecology and Systematics*, 1983, 14:35–55).

sites. And intriguingly, the number of eggs produced by adult female fruit flies is increased by exposure to alcohol vapor, at least for a desert fruit fly species that feeds on rotting cactus tissue. This important measure of reproductive fitness is up to one hundred times higher relative to exposure to water vapor alone. Unfortunately, the physiological mechanisms by which alcohol yields such dramatic effects have never been identified. One possibility, of course, is that the microbicidal action of alcohol helps to fight off bacterial infections, either in the form of infections within the flies or microbial growth in the culture medium that the flies are feeding on. In these ways, the overall health of fruit flies, including their reproductive capacity, could be enhanced relative to zero exposure or much higher levels of alcohol. Natural selection acts directly on reproductive output, so low-level exposure will have real long-term evolutionary consequences.

But how general are these hormetic effects of alcohol? Except for one study showing comparable results with caged rats, we have no data for any other laboratory vertebrate. Very low concentrations of alcohol (e.g., 0.1%) will double the life span of nematode worms under conditions of starvation stress, but these animals are genetically very distant from humans. However, medical observations going back to the 1920s suggested reduced mortality for moderate drinkers relative to either abstainers or heavy drinkers. Systematic work on this important question, however, began only in the 1970s, when the American cardiologist Art Klatsky became increasingly aware of the important role of alcohol in human health. Using large databases, he was able to document statistically a substantial reduction in the incidence of heart attacks. This change was correlated with moderate levels of alcohol consumption. Because heart attacks are a major cause of mortality in modern industrialized populations, any mitigation of their underlying risk factors will have important consequences for the human life span. The qualitative observation that chronic alcohol consumption influences heart physiology actually goes back to 1786, and a variety of observations over the subsequent centuries have amplified and clarified these effects. Epidemiologists today work with sample sizes of tens of thousands of people studied over several decades and are thus able to tease apart possibly confounding social and demographic variables.

To date, an impressive body of evidence has accumulated linking moderate drinking to enhanced overall health. Both abstainers and heavy drinkers experience an increased risk of death relative to moderate drinkers (defined as those who consume one to three standard drinks daily, depending on age and sex; see figure 3). These studies have been replicated by many different research groups working in different (albeit always industrialized) countries, and they spectacularly confirm the initial findings by Art Klatsky. Through the use of detailed survey data and very large sample sizes, numerous socioeconomic factors have also been carefully examined and eliminated as covarying factors that might cause these results (although the effects of ethnicity have yet to be fully elucidated). Moderate drinking can be beneficial in a variety of ways, although one important effect appears to be reduced formation of atherosclerotic plaques within coronary arteries, which otherwise serve to induce heart attacks. Alcohol molecules seem to impede cholesterol deposition and plaque formation within these blood vessels. Moreover, possible antimicrobial action of alcohol relative to various infectious pathogens implicated in atherosclerosis (including

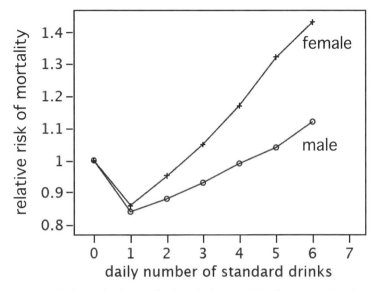

Figure 3. Relative risk of mortality in relation to alcohol consumption by modern humans (modified from Augusto Di Castelnuovo et al., "Alcohol dosing and total mortality in men and women: an updated meta-analysis of 34 prospective studies," *Archives of Internal Medicine*, 2006, 166:2437–2445).

some microbes in our gut) cannot be excluded. Alcohol may thus have a variety of effects in the human body, which together contribute to our longevity in a broad-scale statistical perspective. Missing in all of this research is a prospective experimental study involving randomized treatments of alcohol exposure for large populations; such work would obviously be very difficult to carry out over a decades-long study period. Current cardiovascular benefits of alcohol, as enjoyed by post-reproductive individuals living well into old age, may also not have been of relevance during most of human evolution, when our life spans were much shorter.

Importantly, a number of second-order health effects of alcohol consumption have also been documented. Red wine was initially thought to be more effective than white wine in enhancing cardiovascular fitness, but it turns out the beneficial effects of low-level alcohol consumption are independent of the beverage of choice. It is instead the alcohol alone that has the major protective effect, and up to a point the benefits increase with the rate of consumption. Some compounds in red wine (i.e., polyphenols) may have supplemental effects, but epidemiological studies suggest that these secondary influences are small relative to the physiological impact of alcohol alone. Drinking alcohol with food may also confer greater advantages, possibly because its antimicrobial properties are most effective when it acts not just on food, but also on its inevitable bacterial inhabitants. We all have low-grade bacterial infections in our mouths (as well as in various locations throughout our bodies), and a general system-wide effect of alcohol may be to knock these back to levels more manageable by our immune system. Finally, binge drinking cannot be used to obtain an effect equivalent to consuming the same total amount of alcohol over multiple days. It is, instead, consistent but low-level drinking that conveys the greatest benefit.

A general recommendation of one to three drinks a day, given the protection against cardiovascular risk that would ensue statistically for the drinking population, might then seem to be a reasonable health advisory. This is not necessarily the case, however, and important restrictions pertain to these results. For one, the study populations in question have typically been either from Western Europe or North America (i.e., representatives of industrialized societies), with limited sampling of the genetic diversity present within these populations (chapter 6). Variability in human responses to alcohol, both at the level of enzymatic degradation and of addictive responses, is high and not well understood at the genetic level. A blanket recommendation for drinking behavior cannot be appropriate for all individuals and will vary according to known risk factors for any given individual. The most obvious variable here is sex—males typically drink more than females but also derive the greatest protective benefits from alcohol, albeit at somewhat higher levels of consumption. Individual drinking behavior can also be highly variable through time (and, needless to say, this is

Plate 1. Assortment of rainforest fruits from Barro Colorado Island, Republic of Panama. (Photo by Christian Ziegler.)

Plate 2. The palm *Astrocaryum standleyanum* in the rainforest of Barro Colorado Island.

Plate 3. Fruits of varying ripeness on an infructescence of the rubiaceous shrub *Psychotria limonensis* (Barro Colorado Island). The color progression ranges from unripe green fruits to yellow, orange, and fully ripe red fruits; note also the contrast of ripe fruits against the green foliar background.

Plate 4. Extrafloral nectary (in center) on a Neotropical shrub (*Inga* sp.), with attending ants (*Dolichoderus bispinosus*). (Photo by Phil DeVries.)

Plate 5. A Neotropical fruit-feeding butterfly (*Dulcedo polita*) feeding on a fallen hog plum (*Spondias mombin*). (Photo by Phil DeVries.)

Plate 6. Fruit flies on naturally fallen figs (*Ficus insipida*) on Barro Colorado Island. Fruits are approximately 25 millimeters in diameter; note the white fungal growth on the leftmost fruit.

Plate 7. Ripe fruits of *Astrocaryum standleyanum* on the forest floor, Barro Colorado Island.

Plate 8. An eastern chimpanzee (*Pan troglodytes schweinfurthii*) smelling fig fruit (*Ficus sansibarica*). (Photo by Alain Houle.)

Plate 9. Eastern chimpanzee (*Pan troglodytes schweinfurthii*) and fig fruits (*Ficus sansibarica*) in Kibale National Park, Uganda. (Photo by Alain Houle.)

Plate 10. Supermarket display of alcoholic beverages in Berkeley, California. Compare the colors and diversity (only a small subset of which is depicted here) to those in plate 1.

Plate 11. The New World phyllostomid great fruit-eating bat (*Artibeus lituratus*) removing a fig (*Ficus insipida*) from its infructescence. (Photo by Christian Ziegler.)

Plate 12. Bonobo (*Pan paniscus*) eating a liana fruit in Salonga National Park, Democratic Republic of the Congo. (Photo by Christian Ziegler.)

typically under-reported in survey responses). In turn, epidemiological findings use the concept of the "standard drink" (containing about fourteen grams of alcohol) to try to normalize patterns of consumption. The sociology of drinking behavior is an active field of alcohol research, and one that has important public health consequences for both the positive and negative aspects of drinking behavior. Chapter 6 details many of the deleterious consequences of excessive drinking.

It is also clear that a general recommendation to just "drink more" could easily be misconstrued by the general public. An increase from zero or low levels of alcohol consumption up to the levels of moderate drinking, for example, might also increase the likelihood of subsequent transitioning to excessive or even dangerous consumption. In this context, it is perhaps best to heed the Japanese proverb that a little sake is the best of all medicines. Drinking in moderation is an appropriately simple recommendation, but it also begs the question of how to define moderate. Instead, it is perhaps best to consult the primary medical literature directly, to talk with one's doctor about the issue, and to proceed carefully no matter what choices may be made. Given that the potential consequences of excessive drinking are so dire, a conservative (or dare I say, sobering) perspective on the current state of scientific knowledge is called for.

What are we to make from an evolutionary perspective of these findings that low-level drinking is generally beneficial to modern humans? The theory of hormesis suggests that behavioral and physiological responses towards particular compounds should vary according to relative availability and predictability in the diet. If regular exposure to low concentrations of alcohol is an inevitable consequence of diet (e.g., via daily ingestion), then selection will favor the evolution of metabolic adaptations that maximize physiological benefit and minimize any cost. This argument would, however, apply only to those alcohol concentrations historically encountered by frugivorous animals. Exposure to much higher concentrations of a hormetic substance would, by contrast, be less adaptive and could induce negative responses. Today, low-level alcohol consumption by humans can be advantageous in reducing the risk of heart attack (at least in industrialized societies), and potentially in other measures of overall health. These effects are maximized at fairly low levels of exposure. The potential for abnormal and evolutionarily unnatural levels of consumption, however, is also present given the almost unlimited and cheap availability of alcohol in most modern societies. This is the double-edged sword of hormesis that has tragic consequences for alcoholics (chapter 6).

Most of the time, animals in the real world will never be exposed to large and potentially damaging amounts of alcohol. Instead, the low levels typical of fermenting fruit simply elicit useful behavioral and physiological responses that can be refined over many generations, through the process of natural selection. From this perspective, it is not surprising that alcohol in small amounts actually benefits both animal and human health. These effects are fairly well studied in fruit flies but are likely widespread among fruit-eating mammals and birds as well. Inebriation per se is a very unlikely outcome in the wild, but the alcohol molecule is a familiar companion at mealtimes for many insects and vertebrates. Nowhere is this outcome more resonant than when we consider the ancestry of humans over the past tens of millions of years, among the fruit-eating apes and monkeys of tropical rainforests.

CHAPTER FOUR

Aping About in the Forest

Eating, much like breathing and going to the bathroom, comes naturally and often without too much thought. In the industrialized world, we tend to specialize on a core subset of preferred food items, relative to a much broader range of available cuisine. Much of what we eat involves fermented and microbially charged products such as bread, cheese, yogurt, cured meats, and even the coffee beans that we grind up for a daily cup of joe. These and other food preferences derive primarily from what we grew up with and learned to eat as children and adolescents. As the proud father of a two-year-old, I have lots of opportunity to observe her dietary experimentation and other exploratory tasks that are, frankly, reminiscent of those of monkeys in the wild. In fact, we have inherited numerous behaviors and sensory abilities from our simian ancestors that continue to influence our food choices today. This outcome has been well publicized via the much-hyped concept of the Paleolithic diet and the study of hunter-gatherer cultures in recent times.

Nonetheless, the relevant time horizon in such analyses typically extends back only several million years and is thus misleading. Instead, we need to take a deeper-time perspective on the biology and ancestral foods of primates. To do this, we need to look at the evolutionary origins of primates, the relative extent of fruit in their diets over the past tens of millions of years, and the kinds of environments within which our nearest living ancestors, the great apes, evolved. We need to travel to the great equatorial rainforests, the most species-rich terrestrial habitat on the planet today.

BEASTLY BEHAVIOR

When we think of a tropical rainforest, what typically comes to mind are emerald canopies, green riverbanks, and tall shade trees. Secondarily, we might reflect on the presence of multicolored flowers and fruit, but these are actually pretty uncommon against the background of lush foliage. Ripe fruit is often rare and can be difficult to find in forests. It would be much easier simply to feed on leaves alone, but their caloric return is much lower and requires a much larger investment in digestive machinery. And although the tropics are famous for a high taxonomic diversity of shrubs and tree species, the actual density of fruit-bearing plants throughout the forest is low. It can be difficult at any given time to find enough to eat. And of those trees bearing edible fruit, the ripening sequence dictates that the majority will be green for most of their existence. When they do actually ripen, the scramble for action among microbes, insects, and vertebrates is fast, and fruits disappear quickly. Finding ripe fruits in the forest and then consuming them rapidly is thus critical. Bingeing behavior can be advantageous in some contexts. Large warm-blooded animals, including ourselves, have high caloric requirements, and it can be difficult to meet this fundamental physical need on a daily basis.

A major problem, therefore, is to find fruit in a sea of green, but there is also the issue of seasonal availability. Most tropical forests that do not lie at or near the equator are characterized by a dry season that can be months long, during which time far fewer fruit-bearing shrubs and trees are present. Birds and mammals must travel longer distances to find calories, and there is greater competition both within the same species (as in feeding groups of monkeys) and among different vertebrate species for access to fruit. These conditions require that foraging over distances of tens of kilometers be as efficient as possible. Large, long-lived trees with particularly rich crops may be periodically revisited, requiring good memory and spatial recall as animals climb or fly through dense forest. By contrast, when the monsoon or rains hit these seasonal forests, the plants respond with a flush of leaves, flowers, and fruits that extends over many months. Frugivores can thus go from near famine to feast, and selection for efficient foraging schemes is correspondingly relaxed. When there is more fruit available, animals can also afford to be pickier about what they are eating, and the dietary preferences of frugivores are known to vary seasonally. Assessment of fruit quality and suitability for consumption also changes with hunger level, wariness of predators, and local abundance. We similarly modulate our tolerance of food quality according to hunger level, although most of us live well beyond the danger zone of low calorie intake that often characterizes animals in the wild.

A related factor influencing foraging strategy derives from the presence of most tropical fruits high up in trees. Those large birds and mammals that are confined to the forest canopy tend to be more dedicated fruit eaters, whereas those typically living lower down in the forest understory and on the ground are more omnivorous. For non-flying mammals, moving about in the canopy can be particularly challenging compared to walking on the ground. Frequent leaps and jumps within the canopy are punctuated by travel on large branches, by falls as smaller branches break, and by bouncy intervals as compliant plant structures bend under the animal's weight. Three-dimensional vegetation can thus pose life-threatening challenges for a rapidly moving animal. Searching for food must occur in parallel with locomotion, and animals traveling in groups (as many primates do) also must keep track of others, both visually and acoustically. Group foraging brings about a substantial increase in overall search efficiency, but the fundamental problem of locating limited nutritional resources still pertains. Predators such as large snakes and various large cats (e.g., ocelots and jaguars in the New World tropical forests) are also out and about hoping to score a meal, rendering the daily existence of humble fruit eaters quite a challenge.

Given all of these difficulties, what are the actual cues that animals use to find fruit within tropical forests? In general, it is very difficult to study the sensory biology of free-ranging animals, particularly if they are moving through dense forest some distance above the ground. Field biologists working on primates, for example, typically spend years at their study sites simply trying to put together the basic composition of the diet and to analyze information about the social structure of their focal species. But the physiological mechanisms by which primates find their food can be inferred, to some extent, by combining results from laboratory experiments with the kinds of sensory information provided by plants in natural environments. It is no accident that the most obvious of these cues to us is color, particularly when the vivid reds, oranges, and blues that typically conclude the ripening sequence are viewed against green tropical foliage (see plate 2).

Such direct lines of sight are, however, the exception rather than the rule for animals moving in complex three-dimensional plant canopies. When we walk in tropical forests, for example, our ground-based perspective results in a misleading view of what fruit eaters actually experience. Most leaves, branches, and lianas are located well above ground, whereas from our earth-bound vantage we see clutter but also sections of open terrain between the large tree trunks of mature forest. Higher up, the distances over which fruit can be seen in dense forest can be very limited, particularly up in the canopy, where most fruits are located. Some ripe fruits that have already fallen from their tree can be found on the ground, but even these can often be seen only at short range. Fruit-eating birds and mammals forage daily over distances of tens and sometimes hundreds of kilometers, whereas the bright colors of ripe fruits work best only at much shorter range, at least to those species (including many birds and primates) that possess color vision. Visual indicators of ripeness thus serve as only one of many potential cues for finding an appropriate meal.

Ripe fruits also emit a variety of interesting and attractive odors to potential consumers. These odors are particularly evident for large fruit crops, both those hanging in tall tropical trees and those having fallen to the ground (see plate 7). The numerous flavors of fruit, so exaggerated in the abundance of sweet liquors and cocktails generated by humans today, derive from a wide variety of organic compounds. Many of these are volatile, evaporating from the fruit surface and wafting into the surrounding air. Ambient winds then move these molecules through forests, in and out of canopy gaps, and across open terrain, providing a persistent hint of the calorie-rich pulp that lies upwind. The molecules composing a particular fruit odor are taxonomically quite specific, which is why we can tell a mango from a banana or an apple by smell alone. The one major exception to this trend is alcohol, which as discussed in chapter 2 can be found in all fruits colonized by fermenting yeasts. As long as both sugar and these yeasts are present in fruit, some alcohol will be available as a reliable wind-borne message indicating the presence of calories somewhere upwind.

Plumes of alcohol odor driven by the wind can thus be used by many different kinds of animals, including primates, to find ripe fruit. Just as fruit flies fly upwind when they sense alcohol, so too could vertebrate fruit-eaters use this molecule, by virtue of its obligate association with sugars, as an indicator of potential calories. And a huge advantage of alcohol over visual cues is its tendency to waft over long distances, clearly signaling the availability of ripe fruit to distant consumers. Because even light winds characterize most environments, animals often can simply move upwind to find the sources of a preferred odor. The best studied of these responses occur in the context of pheromone communication in many moths, males of which can travel tens of kilometers to find females broadcasting their scent of sexual receptivity. When winds vary in both magnitude and direction, as is the case both horizontally and vertically in natural vegetation, spatially complex odor fields will ensue. Nonetheless, upwind movement upon sensing a desired odor, coupled with sideways motions to find the scent again when the trail is temporarily lost, can be a highly effective searching strategy over long distances. We know from insects that natural selection for these kinds of behaviors can be very strong. And if such strategies can be found in fruit flies, it should be well within the physiological capacity of the vertebrate brain to carry out similar kinds of responses. However, no field data are available to show this directly, given the difficulties of tracking wild animals in three dimensions and simultaneously measuring complex vapor plumes within forest environments.

But can primates actually smell or taste the alcohol molecule? For a long time, it was thought that the answer to this critical question was an emphatic no. Our own experiences as eaters and drinkers notwithstanding, olfactory responses were generally thought to be poorly developed in primates. Some of their neuroanatomical structures that would be devoted to olfaction are reduced in size. And until recently, dietary exposure to alcohol has not been recognized as a natural occurrence. But a series of behavioral and physiological measurements over the last decade showed that various primate species can both smell and taste many of the organic compounds produced by fermentation, including ethanol and a number of other kinds of alcohol molecules. Interestingly, olfactory thresholds for some of the molecules tend to be lower and thus better than those of rodents, which we think of as being well-endowed in terms of smelling capacity. Although positive behavioral responses of primates to alcohol cues have never been studied in the wild, this possibility clearly cannot be excluded a priori. The multiple cues by which apes and monkeys find fruit at a distance are certainly hard to study in the tropical rain forest, but methods have been devised that can be used to test certain cues. For example, artificial jelly fruits can be made with variable sugar and alcohol content and placed on platforms at different heights in the forest. Both animal visitors and removal rates can then be monitored remotely using digital video recorders to evaluate attraction to, and preference for, different alcohol levels. Colors of these artificial fruits can also be manipulated using food dyes. Such experimental approaches would test directly the hypothesis that animal frugivores are preferentially attracted to alcohol-containing fruits, and could be used to study feeding choices in a variety of tropical birds and mammals.

Once having arrived at a cluster of potentially suitable fruits, primates continue to use both color and odor to evaluate ripeness. But they also add in direct tests of texture by squeezing and testing the mostly larger fruits to assess softness and associated ripeness. Remember that fruits relax their physical defenses once ripe, so as to facilitate consumption by an appropriate disperser. Soft and squishy fruits are thus likely to have more available calories, so it is important for primates to squeeze to pick the ripest fruit. And if you clandestinely watch your fellow human beings in the produce section at the supermarket, you will see remarkably similar behaviors. Multicolored visual displays of fruits and vegetables obviously attract our attention at a distance. But once up close, we then pick up and smell fruits, pinch them to assess ripeness, and look for good colors as well as any evidence of rot. Nothing is known about the role of alcohol in mediating such methods of fruit selection, although the smelling of fruit in hand has certainly been observed in wild primates (see plate 8). A stronger scent of alcohol will indicate sugar-rich pulp, so a powerful odor may tip the balance when deciding which among a cluster of fruits to actually eat. For some tropical fruits, considerable time and effort may be required to peel off the husk and to otherwise prepare the pulp for consumption, so it can be important to select only the best ones.

Animals don't have the option of drinking booze before or during a meal, but alcohol naturally resident within ripe fruit may also play an important role in increasing their rate of food consumption. In humans, drinking prior to eating is well-known as the aperitif effect and results in more eating (chapter 5). For primates in the wild, alcohol could have a similar psychoactive effect that promotes faster feeding, which would help in the face of competition for ripe fruits when supply is limited. More time spent eating slowly could potentially result in the loss of calories to other consumers, be they of the same or different species. And there may also be enhanced risk of exposure to predators while animals are focused on selecting and eating the best fruit. Unfortunately, the effects of alcohol on natural feeding rates have never been studied for wild fruit eaters. Could alcohol in fact function as a feeding stimulant? If so, then this effect, along with other pleasure-related correlates of inebriation, may simply be an evolved and historically beneficial outcome deriving from the presence of alcohol within ripe fruit. As modern humans, then, we have simply inherited this sensory bias linking the products of fermentation to caloric gain.

This perspective also raises the important issue of when an animal decides to stop eating relative to how much alcohol has actually been ingested. Typically, the acts of feeding and filling the gut induce physical stretching, which is monitored by receptors in the stomach wall. These sensors ultimately initiate the behavioral decision to stop eating. The positive rewards indicated by alcohol in food can be physiologically relevant up to the point of a full gut, at which time the hunger signal is neurally overridden by the stretch receptors. Further food consumption, and also that of alcohol, is then necessarily inhibited. There may thus be a natural blood-alcohol content that is attained during feeding on fruit but that can never be surpassed because of dietary satiation. For free-ranging primates, we simply do not know the typical amounts of alcohol ingested during a fruit meal. However, natural limits to consumption may be indicated by the blood-alcohol levels reached when these animals voluntarily stop feeding. Given the naturally low concentrations of alcohol within fruit, these levels are likely to be modest, just as are those attained during the consumption of beer, wine, and liquor by the majority of modern humans. Nonetheless, such limited dietary exposure can easily be circumvented today, given the high concentrations and large volumes of low-cost alcohol that are produced industrially (chapter 5). The potential linkages between alcohol and satiation have, unfortunately, been little explored by the addiction research community. In part, this derives from the tendency to study laboratory animals presented with alcohol in liquid form, rather than obligately mixed with food as occurs naturally (chapter 6).

More generally, animals have evolved many different kinds of behavioral strategies to find and rapidly consume fruit. Over the course of tens of millions of years, foraging within species-rich tropical plant communities has required the evolution of diverse sensory capacities, particularly in response to seasonal variation in fruit production and the huge taxonomic range of fruit-producing plants. And unlike the pollination of flowers, which is often characterized by a high degree of specific association between one kind of animal visitor and one floral type, the evolutionary linkage between fruits and consumers is much more diffuse. Broad ecological syndromes can be identified for bird and mammal frugivores, with the former group tending to eat smaller fruits on average. But frugivorous vertebrates overall tend to be fairly generalist consumers with respect to fruit color, flavor, and taxonomic identity. A wide range of behavioral flexibility is correspondingly required to successfully find and identify potentially thousands of different kinds of food items.

To this end, multiple sensory cues are the norm for use in fruit selection, including vision, olfaction, and touch. Such capacities provide for a greater diversity of feeding possibilities given fluctuations in availability and ripeness. Because good fruit can be scarce in tropical forests, and given that considerable benefits result from finding and consuming it before others do, we can safely assume that there have been strong forces of natural selection acting to promote better foraging strategies. That frugivory has been a successful strategy for many primates is well illustrated by consideration of their diets through the ages and into contemporary times. Frugivory remains a winning nutritional strategy for all of our close relatives, from gibbons to chimpanzees. These are the real fruit specialists.

GIBBONS AT THE TABLE

The reconstruction of ancient diets can be a very tricky business. We can't observe directly what animals were eating millions of years ago, but through a combination of different sets of biological data, it is possible to broadly identify feeding strategies. To some degree, the level of dietary specialization can also be characterized. Carnivores, for example, possess teeth that are very different from those of herbivores. Herbivores feeding on grass have morphological specializations not found in fruit feeders. And of the frugivores, those with thinner enamel on the teeth are more likely to specialize on softer and possibly riper fruits. Patterns of microwear on teeth deriving from mineral inclusions (e.g., silica) in food can also be informative. On the supply side, the broader ecological context of where fossils are found can also provide clues about available dietary resources. Plants in wet tropical forests display pollen signatures different from those in temperate-zone grasslands, and this information can be used to infer the availability of particular kinds of seeds, fruits, and leaves. By considering such multiple sources of data, and particularly through the study of fossil teeth, it is possible to develop a broad-brush but nonetheless informative picture of primate diets through time.

Compared to other mammals, primates are, in fact, relative newcomers to the evolutionary stage. The best physical evidence for their origins places them in the tropical forests of the Eocene, about 55 million years ago, although molecular estimates of primate phylogeny suggest earlier origins, around 80 to 90 million years ago. Much later—around 45 million years ago—some primate lineages diversified as fruit-eaters active during the day within the canopy of structurally complex forests. Among other morphological features, primates are characterized by stereoscopic vision (i.e., overlapping right and left visual fields), trichromatic color vision (in Old World monkeys and apes), and the presence of both fingernails and toenails, fascinating structures not found in any other animal group. Both stereoscopic and color vision are advantageous when scanning in three dimensions for food in dense vegetation, and fingernails are useful for removing the husk or skin from fruits (think about peeling an orange, for example). This fruit-based diet continued for tens of millions of years, at which time the early apes (i.e., the hominoids) first appeared. Again, the teeth of these animals are consistent with a diet of soft and presumably ripe fruit. As the apes subsequently diversified into a variety of differently sized species and ecological niches, they continued to eat fruit, suggesting ongoing advantages to this foraging strategy. Flowering plants diversified as well over these millions of years, yielding a profuse number of tropical tree species and numerous types of fleshy fruit. Some consumption by primates of nectar, flowers, and gummy exudates from tree trunks was also likely.

Such dietary patterns, possibly with a shift from softer to harder fruits, persisted until the appearance around 18 million years ago of what are termed the great apes (i.e., the hominids), the evolutionary lineage that ultimately led to modern humans (see figure 4). Although the historical relationships among different hominid species are not well resolved and are literally based on fragmentary fossil evidence in some cases, it is clear that their diets began to diverge from the predominantly fruit-oriented meals of their predecessors. By four million years ago, groups such as the australopithecines and other lineages of the great apes were clearly eating a much broader variety of materials. Fruit consumption cannot be excluded, but roots, tubers, both freshwater and marine shellfish, carcasses, and hunted animals were progressively added to their diets. Once true humans (i.e., bipedal hominid species belonging to the genus *Homo*) emerged around two and a half million years ago, their diets were clearly much more diverse than those of their predecessors. Nonetheless, our frugivorous heritage is best appreciated when we consider the diets of those great apes and other relatives who are still around today.

Chimpanzees are our closest relatives and exemplify a dedicated strategy of feeding on ripe fruit (see plate 9). The timing of the origination of chimps is not entirely resolved but was probably in the region of four million years ago. Today, numerous ecological studies have indicated that each of the two extant species (i.e., the common chimpanzee and the bonobo, both in the genus *Pan*) feed preferentially on energy-rich ripe

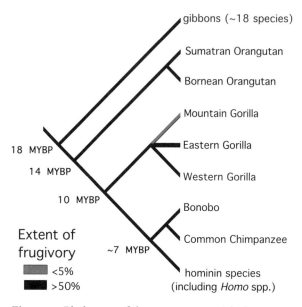

Figure 4. Phylogeny of the extant apes, with the relative extent of frugivory in each group and the most parsimonious reconstruction of diet for ancestral lineages. With the exception of the mountain gorillas, all extant taxa, ancestral lineages, and the lineage leading to the genus *Homo* are primarily fruit eaters. Dates indicate approximate timing of the split between two lineages; MYPB: million years before present.

fruits, which compose over 85% of their diet. But, just as with reconstructing fossil diets, food selection by extant primate species can be variably categorized. Different kinds of assays (i.e., total time spent foraging, estimated percentage of caloric intake, etc.) have been used by different scientists to assess the relative contributions of particular dietary items. As mentioned previously, considerable seasonal as well as geographical variation in diet can be found for any primate studied in the field. Chimps even hunt down and consume monkeys from time to time. Nonetheless, broad dietary categories can be useful to characterize primate species if the subtleties and different methodologies associated with such discrete semantic labels are kept in mind. For chimpanzees, their extent of specialization on fruit is so high that their dietary designation as frugivores is unambiguous.

Moreover, such specialization characterizes most of the other extant apes as well. The genus *Gorilla,* which comprises three distinct evolutionary lineages, split from the other great apes somewhere between eight and eleven million years ago. The two lowland species of *Gorilla* (i.e., eastern and western gorillas) today have fruit-based diets similar to those of chimpanzees and can be found in much the same kinds of tropical forests. The highland lineage of *Gorilla* (which has also been treated taxonomically as a subspecies of the eastern gorilla) is the sole exception to the trend of frugivory among modern ape species. Instead, highland gorillas eat mostly herbaceous vegetation, although some smallish fruits can be seasonally available. Large sugar-rich fruits are in fact mostly absent at elevations greater than 2,000 meters in the tropics because the energy balance of plants is simply insufficient to support the net production of such large amounts of sweet pulp. Butterflies that would otherwise feed on fallen fruit are similarly absent in tropical montane habitats.

Given the ancestral habitat of the great apes in the tropical lowlands of Africa, the diet of highland gorillas can thus be viewed as a secondary adaptation to montane life. By contrast, those in the lowland exemplify a successful dietary strategy based primarily on the fruit of tropical rainforests. Although ground dwellers by virtue of their massive stature, these great animals can subsist by foraging opportunistically on a range of fallen ripe fruits, but also include leaves and other plant products. Representing a deeper split in evolutionary time, the orangutans of Southeast Asia (with two extant species) are also strongly frugivorous, as are the most basal apes, the gibbons (around twenty species). Both orangutans and gibbons spend most of their time in the trees and feed predominantly on ripe fruit. Nearly exclusive or partial frugivory thus characterizes approximately 24 million years of hominoid evolution and virtually all of our close relatives today.

By contrast, dietary diversification within the last four million years clearly characterized the hominids, and particularly the bipedal lineage leading to ourselves. Inclusion of animal fat and protein became increasingly important, although even for modern hunter-gatherer societies these sources of food represent less than 50% of the total diet. With the advent of agriculture around twelve thousand years ago, along with a developing technological capacity for food storage, human diets transformed rapidly relative to those of apes in tropical forests. Nonetheless, we enjoy a deeply rooted fruit-eating heritage courtesy of our ancestors. The extent to which we may have retained neural and behavioral associations of alcohol with the caloric rewards of ripe fruits has, however, never been experimentally evaluated. Particularly useful would be knowledge of olfactory responses to alcohol in a variety of primate species, fruit feeders and otherwise, for comparison with the physiological capacities of modern humans. Similarly, it would be nice to know about the genetic and molecular underpinnings to such sensory abilities among different species of fruit eaters. Most readers of this book will need no convincing that we have a special relationship with alcohol. Deeper tests of the hypothesis, however, will require much more comparative data from our simian relatives who still reside in the Old World tropics. Humans long ago migrated from these regions to dwell on all continents and now live in food environments very different from those of the past. Conceptual similarities of urban life to a jungle notwithstanding, we have constructed entirely new domains for expressing dietary choice.

FORAGING IN THE CONCRETE JUNGLE

Compared to the daily taxonomic variability that characterizes the diets of wild monkeys and apes, the eating habits of most modern humans are much more stereotypical. The fundamental constituents of our body obviously must derive from our choices in food, but most of the time we are on automatic pilot relative to what we eat. We typically consume three major meals a day but often spend little or no time thinking about what we eat. Instead, we stick to what we know and eat only a very limited subset of all possible food items. In many contemporary societies, patterns of food consumption have also been dramatically transformed relative to what they were even a hundred years ago. The taxonomic and nutritional ranges of available food have never been greater in industrialized nations, but so too are the possibilities for excessive consumption of highly processed and chemically manipulated compounds. These modern foodstuffs are very far removed from their natural origins.

Salient among these are the cheap carbohydrates and fats that form the backbone of most of the industrial food products featured in the modern supermarket. It is becoming increasingly clear that high-level, and of course unnatural, exposure to such dietary constituents can damage human health. Premier examples of this outcome include diabetes and obesity, both of which are partially heritable but the occurrence of which is also strongly influenced by environmental factors. The current obesity epidemic in the United States and the increasing incidence of diabetes worldwide are compelling testimony both to the efficiency of modern food production and to our intrinsic behavioral reactions to cheap food. The cascade of negative consequences for our health is ignored in the short term as our sensory biases promote consumption of excessively sugary, salty, and fatty foods, all nicely packaged in colorful and attractive ways courtesy of the advertising industry. Why then do we consume these items at levels far greater than might match historical levels of exposure, even when the long-term consequences can be morbidity, hospitalization, and even premature death?

An emerging field of inquiry termed evolutionary (or Darwinian) medicine can at least partially answer this question. This discipline seeks to apply principles derived from evolutionary theory to major problems of human health. Salient examples include selection for increased antibiotic resistance in microbes, the origins of such metabolic conditions as sickle-cell anemia and lactose intolerance, and the medical consequences of mismatches between our ancestral and modern environments. If any behavior that was historically beneficial is even partially heritable, for example, then its genetic retention into modern times may in fact be disadvantageous if it now promotes extreme or deleterious behavior. Many of us now live in technology-dominated worlds that are widely and sometimes rudely divergent from those of our ancestors, whether those of ten thousand or ten million years ago. At the same time, many features of our neural and sensory biology have been retained. These include abilities that critically underpin survival and fitness, such as awareness of predators, parsing of the often ambiguous cues used in mate selection, and various features of dietary preference. Animals in the wild are energetically challenged, and similarly we all have a persistent drive to acquire calories. But often we don't pay much attention to what we are eating. Instead, choice of diet can be abandoned to an autopilot in our brain that channels the results of millions of years of selection.

This biological drive to meet basic metabolic demands underlying everything else we do in life can directly influence mood and behavior. Most of us get grumpy if we miss a meal, for example. When calories in the environment are sparse, as characterizes most if not all natural habitats, it is critically important to be able to forage efficiently and to consume avidly when possible. Natural selection has acted to ensure that such behaviors are the ones that have persisted genetically over thousands of generations. These otherwise adaptive strategies can go badly wrong, however, when industrial food production lets us indulge every day in cheap calories, often to extremes. Such a mismatch between historical and contemporary dietary environments has led, in the terminology of evolutionary medicine, to diseases of nutritional excess. Uninhibited and sometimes self-destructive consumption of calories in modern food environments simply reflects our innate tendency to behave as if eating more were always a positive thing. Unfortunately for some, the stopping mechanisms just can't deal with the vast amounts of food available today. Such addictions, in other words, may be an accidental but perhaps inevitable consequence of living in the modern world.

Could our responses to alcohol reflect a similar outcome? If natural selection has acted on human ancestors to associate this molecule with nutritional gain, then its psychoactive features may simply promote evolved and once advantageous behaviors that encourage its rapid consumption. These responses then carry over into modern times, but with the important difference that alcohol is now widely available and at much higher concentrations than was ever the case for fermenting fruits. Addiction obviously involves a number of factors supplemental to routine ingestion, but essential to any hypothesis of abuse must be an explanation of what motivates attraction to the molecule in the first place. We must note, however, an important assumption of this hypothesis. Genetically based behaviors promoting low-level but routine alcohol ingestion via frugivory must have been the targets of selection during human evolution. Alcoholism is well known to be partially heritable (chapter 6), which is consistent with the possibility that this disease derives, at least in part, from foraging behaviors that once were adaptive.

Another important prediction from the field of evolutionary medicine is that animals will acquire, through natural selection, the metabolic capacity to take advantage of low-level but routine exposure to otherwise toxic compounds. As discussed in the previous chapter, moderate daily ingestion of alcohol can indeed be beneficial to humans today, yielding substantial reductions in both cardiovascular risk and overall mortality. Intriguingly, genetic variation in the physiological reaction to alcohol, as well as in the susceptibility to alcoholism, can also be identified for different human groups (chapter 6). These results suggest that alcohol exposure among human populations has been historically variable, possibly within the last ten thousand years. The health consequences of alcohol exposure may also be genetically correlated with the capacity to metabolize the molecule. In turn, rapid evolution of associated physiological responses would be evident today in non-random and regionally restricted patterns of variation in associated genes. Together with similar results in fruit flies (chapter 3), such differing levels of exposure, along with the known positive effects of alcohol, would suggest a more general evolutionary outcome. Ultimately, and thanks above all to the fermenting yeasts, we now enjoy physiological benefits of alcohol exposure (such as an enhanced lifespan) that may well characterize all fruit-feeding animals.

To test this hypothesis, it would be necessary to determine the long-term consequences of low-level alcohol consumption for a number of different primate species, for other fruit-eating mammals and birds, and even for nectar-feeding animals that might experience routine exposure to alcohol in the wild. Both the costs and benefits of exposure would have to be measured, and it would be necessary to monitor individuals over their entire life span so as to determine effects on longevity. This requirement would be particularly difficult to carry out for long-lived primates but would test a key prediction of the evolutionary hypothesis linking alcohol exposure to overall health. If indeed this is a general result not specific to human beings, we may then have confidence in the broader relevance of the drunken monkey hypothesis for the natural biology of alcohol consumption.

The drunk on the street, of course, doesn't really care about the evolutionary origins of attraction to alcohol. Via an ecological linkage with fruit-based calories, natural selection has engendered impressive reward functions that kick in whenever we drink. The innate craving for booze has been documented in virtually all human cultures, and a variety of health benefits derive from its routine low-level consumption. But we've also figured out how to produce alcohol at amounts and concentrations much higher than those found in natural situations. This mismatch relative to historical availability can unfortunately lead to excessive levels of consumption and abuse. The attraction to drink nonetheless remains powerful and must have accelerated with the origins of agriculture and intentional fermentation. The yeasts are still the same, but via simple technologies we have dramatically enhanced our capacity to provide alcohol and to get smashed.

CHAPTER FIVE

A First-Rate Molecule

Alcohol is a popular molecule in many human societies. We drink either alone or together in a wide range of circumstances: at home, in restaurants and bars, at parties, and sometimes in religious ceremonies. The amounts of booze used for all of these purposes are huge. The alcoholic beverage industry in the United States produces billions of liters of beer, wine, and spirits annually. (Ethanol production for varied industrial processes and for transportation fuel is even greater.) Equally of interest are the remarkable patterns of cultural variability in behavioral and social responses to alcohol as recorded by anthropologists over the centuries. Not surprisingly, much of this rich literature describes associations between drink and food. But just when and how did alcohol become so important to modern humans? Who first had the great idea to intentionally ferment fruits and grains?

BREWING UP A STORM

Historians and biologists alike would love to know when and where early humans first gained the ability to produce alcohol in any substantial quantity. Unfortunately, these events cannot be uniquely fixed in time, although we can certainly speculate as to various ecological factors that may have been involved. The first important observation is that fermentation is a natural occurrence given the simultaneous presence of yeasts and sugars in ripe fruit. Clusters of hanging fruit, together with fallen crops on the ground, would present abundant opportunities in both tropical and temperate-zone regions for the spontaneous production of alcohol. Certainly early humans, along with numerous other animals, could find and consume these resources. The second point is that anticipation of the appearance of ripe fruit, which requires both memory of the locations of appropriate plants and a sense of seasonal timing, could potentially lead to a long-term association between ripe fruit and the positive psychoactive effects deriving from its consumption. Any decision to actively gather and ferment fruits would necessarily rely on this connection, although when and where this first happened must remain speculative. Diluted honey will also ferment, and this avenue for the initial observation of naturally occurring alcohol, perhaps then emulated by enterprising humans, cannot be excluded.

An entertaining Chinese story about fermentation nicely illustrates the human tendency, in the absence of supporting evidence, to project modern behaviors concerning alcohol deeper into our biological past. In 1985, the newspaper *Anhui Daily* reported that monkeys in the Huangshan region (i.e., the Yellow Mountains) were caching fruit in rock crevices with the express intent of producing alcohol. Ostensibly, the monkeys were then returning to the cached fruit several months later and feasting on the alcohol-laden product. Such an intentional behavior would be truly remarkable for a non-human primate; it's much likelier that the animals in question were simply feeding on various fruit crops, some of which may have fermented while lying on the ground. Although many birds and mammals do indeed store and later retrieve different kinds of plant materials, this behavior is typically confined to non-fermentable seeds and nuts that can last a long time in natural environments. Rapid rates of fruit decomposition and ensuing evaporation of alcohol would, in any event, prevent any buildup of spirits over more than several hours. Nonetheless, the natural fermentation of wild-collected fruits is well-known today among the Chinese to yield what is called *hou jiu,* or monkey wine. A similar term (*saru zake*) is used in Japan relative to either intentional or accidental fermentation of fruits accumulated by animals within tree hollows. In no case, however, has this outcome ever been scientifically documented or studied in any systematic sense.

In contrast to the monkeys, anatomically modern humans have been around only for several hundred thousand years, and our experimentation with alcohol production is much more recent. In archaeological contexts, direct evidence for intentional fermentation dates only to the Neolithic, that is, to about 10,000 BCE and later. Such documentation, most of which has been conducted by the biomolecular archaeologist Patrick McGovern and colleagues over the last two decades, is based on the identification within ancient storage vessels of chemical residues that are uniquely associated with beer and wine production. This approach uses modern methods of analytical chemistry to identify those varied compounds that can derive only from the fermentation of diverse plant products, including fruits and grains but also honey, nectar, and even tree resins. And although the origins of large-scale fermentation have been traditionally thought to lie in Mesopotamia, exciting discoveries reported in 2004 now place the earliest known site (dated at about 7000 BCE) in the northern Chinese province of Henan. Jars from this site contained a residue which, upon chemical analysis, indicated a variety of fermentables based on their molecular signatures, including grapes and/or hawthorn fruit, rice, and honey.

By contrast, the oldest archaeological evidence for intentional fermentation in the Middle East is thousands of years later (about 5400 BCE), at a site in the Zagros Mountains of western Iran. Here, jar residues containing tartaric acid are consistent with the presence of wine made from the Eurasian grape. This plant grows wild in the region but may also have been cultivated. In addition, other jars at this site contained chemical residues of fermented barley, indicating production of beer. The production of wine, beer, mead, and mixed fermented beverages has now been identified at other later Neolithic sites throughout the Mediterranean and the Middle East. The residues within pottery at many of these locations are also characterized by the presence of multiple plant additives, including resins and aromatic compounds, which serve to preserve and flavor alcohol mixtures. As the millennia progressed, the technical skills of grain, fruit, and natural product preparation, yeast propagation, and careful handling and storage of the final fermented product all became progressively more refined.

The origins of agriculture, and indirectly of large-scale civilization, were thus inextricably bound up with the early practices of alcohol production. And we can't rule out the possibility that intentional fermentation of wild crops actually preceded more targeted agricultural efforts at growing these plants. Broadly speaking, the period of the Neolithic corresponds with the origins of plant cultivation, farming, and the storage of gathered crops at a number of Old World sites of incipient civilization. Domestication of the principal cereals around the world (i.e., barley, corn, millet, sorghum, rice, and wheat) must have proceeded in parallel with their intentional fermentation to yield alcohol, fueling human metabolism, social events, and broader economic activity alike. Concurrently, pottery and other types of vessels to store water, grains, and fermented products became commonplace features of society. Drinking to relax and to mitigate psychologically the hardships of everyday life must also have become a regular if not almost ritualized feature of society. Brewing of beer, in particular, and associated feasting may have contributed synergistically to cereal cultivation and social organization.

Liquid alcohol was not necessarily the only goal of controlled fermentation, as this process can also be a useful means of preserving food. For example, if the appropriate kinds of bacteria are present in addition to fermenting yeasts, acidic molecules will be produced that serve to stabilize and preserve otherwise degradable foodstuffs (as in cheese, sour milk, yogurt, sauerkraut, miso, and kimchi). Additional flavors and aromas also derive from such fermentations, as well as enhanced digestibility (and decreased or negligible cooking time) through breakdown of proteins and complex carbohydrates. Similarly, the presence of alcohol will extend the shelf life of perishable fruits and of gruels made from grain or other starchy products. Wheat and barley, both of which were early targets of domestication in human prehistory, can be used to create beer as well as bread. With the advent of organized agriculture and seed retention for later planting, the storage of dry grains promoted further innovations in pottery design and production. In parallel, the large-scale production of boozy suspensions from fermented wheat seeds and other plant starches was almost inevitable. The roots and stalks of some plants (e.g., sugar cane) also contain fermentable sugars that, following maceration and mechanical extraction, are suitable for alcohol production.

The incipient crafts of brewing, wine-making, and other modes of alcohol production were in full swing, in other words, as early civilizations arose throughout the Old World. But unlike the fermentation of ripe fruits with abundant simple sugars, the use of grains such as barley, sorghum, millet, and rice requires an additional biochemical agent to break down complex carbohydrates into the smaller sugar molecules used by yeast cells. This process (technically known as saccharification) can be accomplished through the use of specialized molds, as is done today for many rice brews in East Asia. It is also possible to recruit naturally occurring enzymes within sprouting seeds, as in the malting agents found in germinating barley, or to use the amylase enzymes present in human saliva. In all such approaches, careful attention must be paid to temperature and other factors influencing yeast growth as the fermenting liquid matures. For our Neolithic forebears, repeated experiments in alcohol production, including tweaking of additives and repeated tastings as fermentation proceeded, would have resulted in an essentially Darwinian process of positive selection for higher yields. Such knowledge would then reinforce the essential database used for agricultural production and would quickly propagate within an emerging society. Also, the psychoactive effects that come into play during drinking would presumably enhance the personal reputations of those who had mastered the art of fermentation. Positive cultural feedback, in other words, would tighten the linkage between domesticated crops, social networks, and alcohol production.

An interesting and unresolved question is whether the East Asian and Middle Eastern practices of controlled fermentation represent independent geographical origins for this cultural innovation. Alternatively, diffusion of knowledge from a single region of alcohol production could have subsequently facilitated widespread use at multiple sites in the Old World. Archeological evidence may never be sufficient to resolve this issue, but it is abundantly clear that most Neolithic efforts in fermentation left no trace whatsoever. It is also notable that only limited physical evidence for early fermentations has been found in the New World, although indigenous production of alcoholic beverages (e.g., *chicha,* the Andean drink made of masticated corn kernels or other carbohydrate-rich plant products) is widespread. The important point is that fermentation is very easy to accomplish if a sugar-rich substrate can be either harvested from the wild or obtained from the preparation of cultivated grains. Just as incremental steps in the domestication of multiple crop plants would have been carried out innumerable times at different locations, so too must experiments in alcohol production have been widespread. And for thousands of years following the Neolithic revolution, movement of both individuals and larger human groups would have spread these features of civilization, including agriculture and controlled fermentation, throughout the Mediterranean and into the far reaches of Eurasia. Such practices were accordingly widespread in the post-Neolithic world.

Although the ancient origins of grain and fruit fermentation can be discerned through the careful study of archaeological remains, it is also clear that the ensuing alcoholic beverages were limited to fairly low concentrations (i.e., no more than 15%, and probably much lower). When and where did the higher concentrations that we associate with spirits and concentrated liquor arise? Again, archaeology can tell us something about the technologies initially employed to enhance the levels of alcohol. Once a fermentation yields at least some alcohol (along with a lot of water mixed in), specific chemical methods are necessary to obtain mixtures with a higher concentration. Two primary means were developed by humans to this end. The first of these is so-called freeze distillation, which relies on the lower freezing point of alcohol relative to water. By externally applying ice and thus reducing the temperature of a brew, the pure water freezes out while the alcohol remains in liquid form on the surface. The fluid on top can then be scraped or siphoned off to produce a high-concentration alcoholic drink. This clever method was known literally as "frozen-out wine" in the Tang dynasty of China (starting around 600 CE) but may have been used in both East and Central Asia several hundred years earlier. Freezing out of alcohol was also common in colonial North America for the production of applejack from fermented cider. Today, the method is used by modern industry to produce ice beer and ice wine. Freeze distillation is, however, time-consuming to carry out and inevitably mixes the alcohol with some icy water, reducing total yield.

By contrast, a much more efficient means of concentrating alcohol is that of vapor distillation, an approach used today by moonshiners and industrial producers alike. Because water and alcohol vary not just in freezing point but also in boiling point, it is possible to slowly heat up fermented mixtures and thus to drive off the alcohol alone in gaseous form. These molecules are then passed through a cooling column, resulting in condensation of a higher-concentration alcoholic liquid that drips into a collector. Repeated bouts of distillation of this enhanced product result in progressively better, more concentrated alcoholic yields. As a chemical process to concentrate different kinds of liquid substances, vapor distillation reputedly dates to the Greek alchemists active in Alexandria around 100 CE. Its specific application to refine alcohol may have occurred around the same time in China, given the existence of two bronze stills dating from the Eastern Han dynasty (25–220 CE). Abundant evidence for distilled spirits can be found from about 700 CE onwards in China, where they were and remain known as "burnt wine."

As with the origins of intentional fermentation, the actual site or sites for the initial innovation of this process are unknown, as are subsequent patterns of cultural transmission throughout the Eurasian continent.

Alcohol is, of course, a useful solvent for extracting plant compounds for medicinal purposes, and much of the geographical spread of vapor distillation may have also been associated with this particular application. The use of low-concentration alcohol solutions as a disinfectant may well have preceded the origins of distillation, and of course any consumption of alcohol could potentially enhance life span and improve overall health (chapter 3). The use of distilled spirits in botanical tinctures clearly would have added to this medicinal repertoire. But remarkably, distillation of alcohol in western Europe is reliably dated only to the early Middle Ages, in Italy. By the late Middle Ages, the production of distilled spirits for the specific purposes of inebriation was in full swing. The word "alcohol," in fact, derives from the Arabic, reflecting its origin in basic chemistry as practiced in the medieval Middle East. Its first recorded use in the English language (in 1543, according to the *Oxford English Dictionary*) was to indicate a heated extract from chemical substances, and only centuries later is the word specifically used to refer to the inebriating products of fermentation. But once the process was mastered, high-concentration alcohol was obviously here to stay, tapping deeply into human physiology and psyche alike. This outcome is even more impressive when we consider the relative novelty both of intentional fermentation and of distillation in human history. Relative to the appearance of anatomically modern humans about 200,000 years ago, our cultural exposure to substantial volumes of beer, wine, and stronger alcoholic mixtures is strikingly recent.

DRINKS ALL AROUND!

Today, many industrialized societies would seem to be awash in alcohol. Consumption is clearly demand- rather than supply-limited, with an astonishing diversity of cheap beer, wine, and distilled spirits readily available at liquor stores and supermarkets alike (see plate 10). Given the low-cost production of carbohydrates through industrialized agriculture, and the fact that the work of chemical transformation to alcohol is performed gratis by yeast cells, we can perhaps both revel in and mourn the fact that many kinds of booze cost less than high-end mineral water. And the sheer amounts of alcoholic beverages produced and consumed annually are staggering. On average, a citizen of the United States drinks about nine liters of pure alcohol every year. About half of this is in the form of beer, which has a fairly low alcohol content (4 to 8%, depending on the brew), whereas the volumes of alcohol in typical servings of wine and mixed drinks are two to four times higher. And pure shots of liquor lie in the 40 to 60% range, although there is considerable variation in the total amount of alcohol found in any individual drink. Nonetheless, the vast majority of alcohol consumed by humans is taken at concentrations much higher than those found within naturally fermenting fruit.

But as we all know, rates of alcohol consumption can be highly variable among individuals. Some of us monitor our intake carefully, and others seem to mindlessly quaff whatever comes to hand. And even at the level of nation-states, systematic differences in drinking behavior are striking. Countries with predominantly Muslim populations, for example, are characterized by much lower reported rates of per capita alcohol consumption relative to other cultures. Many high-level consumers, by contrast, tend to be in north and central European countries, including Russia. Consumption rates are also impressive throughout Africa, where both religious use of fermented substrates (e.g., bananas, honey, sorghum, and millet) as well as their daily consumption to meet caloric demands are commonplace. Much published data on consumption rates should be viewed cautiously, however, as cultural prohibitions and other factors likely preclude accurate quantification of drinking behavior. Individuals who drink regularly also routinely underestimate the actual volumes of alcohol that they consume. According to World Health Organization estimates from 2005 for both recorded and unrecorded levels of consumption, the highest per capita rates are found in Moldova, whereas Yemen is reportedly characterized by almost no drinking whatsoever. Also informative are national rates of industrial-scale production of alcohol-containing drinks. Not surprisingly, France, Italy, and Spain lead in the total volume of wine produced annually. Relative to alcoholic beverages, however, much greater volumes of alcohol are produced worldwide for various chemical and industrial purposes. Because alcohol retains most of the energy of the sugar molecules from which it is derived, it can also potentially serve not just as a means of inebriation, but also as a liquid fuel. In Brazil, for example, about 20% of all automobile fuel is formulated from the alcohol derived from fermentation of sugar cane.

Because the energy content of alcohol molecules is so high, the number of calories consumed in booze by heavy drinkers can represent a significant fraction of their daily energy intake. A remarkably high threshold of 50% of total caloric expenditure derived from alcohol has even been used as one (of many) definitions of alcoholism (chapter 6). Clearly, such a cutoff would indicate excessive consumption by any standard. At lower but still substantial levels of drinking, the beer gut phenomenon serves as ample testimony to the calorie load imposed by chronic consumption. More typically, most who regularly consume alcoholic beverages derive from 2 to 10% of their total metabolic calories from their drinking. However, both the alcohol content of drinks and the timing of consumption will influence long-term medical consequences of exposure. The physiological effects on the body of binge drinking, for example, are very different from consuming the same amount of alcohol spaced out evenly over multiple days. Regular drinkers also tend to consume more alcohol as they age, contributing significantly to total calorie intake and associated weight gain over the decades. Men tend to drink more, on average, than do women. It is important, however, to note that substantial fractions of the United States population (approximately 30%) and of the human population globally (roughly 50%, mostly in Africa, the Middle East, the Indian subcontinent, and parts of East Asia) are reported to abstain from alcohol. Comprehensive information on drinking behavior throughout these regions is sorely lacking.

For those of us who do drink, the behavioral effects of exposure to alcohol ultimately derive from the action of individual drug molecules on our central nervous systems. The action begins when imbibed drinks first travel through various conduits of our anatomy, including the lips, stomach, and intestines. Molecules of alcohol are then absorbed onto and also transit through the internal linings of the digestive and then the circulatory system before exerting their neurobiological influence. Epithelial linings of our mouth first pick up a small fraction of the alcohol within drinks. But most booze passes quickly down the esophagus into the stomach and small intestine, where drinks and meals commingle. Most alcohol molecules are then absorbed by our gut lining and are picked up by capillaries and moved rapidly to all body tissues via the blood. The enzymes that degrade alcohol can be found throughout our body, although most are found within the liver (some 10% of which is dedicated to this task). But prior to this point, the relatively small alcohol molecules circulating in the blood have ample opportunity first to enter our head, and then to transit the blood-brain barrier (which is otherwise impermeable to larger molecules). Alcohol then rapidly circulates within our brain. At any point thereafter, the molecules can diffuse quickly onto and inside any of the neurons that, in aggregate, compose our brain and the majority of our internal control mechanisms.

In contrast to many other pharmacological agents that exhibit tight binding to specific receptor molecules on neurons, alcohol is a very broad-acting drug. Drunkenness and the various other psychoactive effects induced by the molecule cannot be pinpointed either to specific regions of the brain or to particular types of neurons. Instead, a variety of ion receptors and chemical pathways are influenced by alcohol. Many of the latter are reward pathways (especially those involving the neurotransmitter dopamine) which have been implicated in the development of addictive responses to many types of drugs. As a consequence, the overt behavioral effect of alcohol is initially to induce positive feelings and to reduce inhibition and anxiety. Enhanced social interaction can, of course, be another major consequence (see below). These changes depend, however, on the amount and rate of alcohol consumed, and they occur primarily at low blood-alcohol levels. Some of the other effects of initial exposure are equally impressive. Alcohol acts to increase the permeability to water of filtering structures in our kidneys, increasing rates of urine production and visits to the bathroom. At higher levels of alcohol consumption, however, things can go badly wrong. Depressive and other adverse psychoactive and physiological outcomes ensue, including impaired judgment, slurred speech, and loss of motor control. Sufficiently high alcohol consumption induces nausea and ultimately loss of consciousness, sometimes leading to death. One only need scan the headlines of local news sources to learn how routine such tragic outcomes are.

The dosage-dependent effects of alcohol are paralleled by our widely varying responses to this inebriating substance. The positive feelings associated with low-level consumption are readily and widely appreciated, whereas the negative consequences associated with substantially higher levels of exposure are attained much less frequently and by far fewer drinkers. Just as with the U-shaped curve relating human mortality to alcohol consumption (chapter 3 and figure 3), many other effects of exposure depend in complex ways on both the timing and volume of consumed drink, as well as on the physiology and personality of the individual imbiber. This empirical outcome makes scientific recommendations as to the appropriate levels of drinking across entire populations very difficult to establish. Further complicating the picture, many people who drink regularly and safely do so in a social setting. Our responses to alcohol can depend on who is around us, with whom we are interacting, and of course the speed at which we drink. Drinking behavior, in other words, derives in part from the conviviality and perceived responses of others to our own inebriation. And it would indeed be surprising if this result were otherwise, given the importance of social interaction for virtually all primate species, including ourselves. Technically, this effect has been termed the situational specificity of tolerance, reflecting the importance of the surrounding social environment for drinking behavior. As we will see in the clinical phenomenon of alcoholism, addiction to alcohol and its ensuing negative consequences often reflect factors that lie well beyond one's immediate physiological responses to the molecule.

Such complex outcomes are similarly reflected in the cultural diversity of drinking behaviors. Over the last century or so, anthropologists and others have had great fun visiting different cultural groups around the planet, tossing back brews with the locals, and recording details of the event for academic posterity. Often termed fieldwork (and probably funded by others), this line of research must be highly enjoyable and rewarding! One interesting finding across such studies is that a very large number of different sugar-containing substances are used for fermentation, including such unlikely substrates as palm sap, bananas, and potatoes. Similarly, the timing and contexts of alcohol consumption are also highly variable. Dinner parties, ceremonial slaughters, and weddings and funerals are routinely characterized by public consumption (and appreciation) of alcoholic beverages. Social context obviously plays a major role in human drinking behavior, as does the use of alcohol to mediate and, in some cases, to facilitate expression of emotions and beliefs. Particularly at higher levels of consumption, however, cultural expectations usually mold our responses to alcohol. Nobody wants to violate deep-rooted social protocols, even when drunk. By way of example, observe the inebriated Japanese businessman in a Tokyo subway station, fearlessly urinating onto the tracks but otherwise behaving decorously in spite of a high blood-alcohol level attained during a dinner with important colleagues.

In fact, drinking behaviors today often involve the consumption of different food items as well. Social events obviously facilitate group preparation of meal precursors and their cooking, but the simultaneous consumption of alcohol is equally important for many cultures. As suggested elsewhere in this book, this outcome may simply reflect an evolutionary linkage between dietary calories and the derived alcoholic products of fermentation. Via technology, modern humans have managed to physically decouple the ingestion of solid food from the intake of liquid alcohol. But at the same time, eating and drinking remain intimately connected in many modern societies. Nowhere is this effect more evident than in the almost ritualistic association between alcohol and cuisine as practiced in restaurants worldwide on a nightly basis.

THE WINE LIST, PLEASE

Along with the din of conversation, utensils, and tableware, the splash of alcohol-containing beverages and the clinking of glasses are sounds we routinely associate with dining out. Not surprisingly, restaurants derive large fractions of their profits from the sale of alcoholic beverages (compare, for example, the prices of equivalent volumes of the same beer or wine for sale in supermarkets and at your local restaurant or diner). And the range of available drinks staggers the imagination, ranging from straight-up shots at takeout shacks in the American South to the literally thousands of vintage wines available at high-end Manhattan restaurants. Between these extremes lies the routine consumption of mostly cheap beer, wine, and cocktails. This last category of drinks is particularly interesting in that fruit-based liqueurs usually form their base, and their sugar content is markedly high. Socially acquired preferences must contribute in part to the tendency of humans to consume such fanciful concoctions, but some of these effects must also derive from the evolutionary association between fruits, sugars, and alcohol. Based on our immediate sensory responses, the memories of certain flavors, and the anticipation of pure pleasure, we mix and often obsessively match certain kinds of alcohols, herbal extracts (e.g., bitters), and fruit-derived sugars.

Similarly, we might try to understand why humans are so obsessed with the apparently innumerable varieties of wine, beer, and spirits produced worldwide. An ancestral sensory bias hinting at beneficial calories might well act as the primary motivating factor to consume alcohol. But what possible advantages could derive from discrimination of the other numerous by-products of fermentation, including the many longer-chained alcohols and literally thousands of additional organic compounds that contribute to flavor and aroma? One explanation might simply be that our taste preference for alcohol, along with other kinds of choices in food, reflect factors of both nature and nurture. We may inherit patterns of genetic variation in our physiological ability to sense and enjoy a number of nutritional compounds, including fermented beverages. Drinking habits are then further modified through youth and adulthood because of what we are habitually provided, consume, and ultimately identify as preferred beverages. In other words, our senses of taste and smell are individually unique because of combined genetic and environmental influences. Preferences for different kinds of alcoholic beverages would then be just as likely to mirror contemporary cultural diversity as our evolutionary past.

Equally relevant, however, is the fact that tropical fruits are taxonomically very diverse and present to frugivores many different kinds of flavors, odors, yeasts, and chemical products of fermentation. The ability to sense and choose among such molecular features may have been useful to human ancestors to discern ripeness and energetic value, and this physiological capacity has perhaps been carried over into modern times. Also, a detailed knowledge of brews, fermentations, and vintages (not to mention the wonderful French concept of terroir) carries social weight in many contexts today. Conversancy with the esoterica of wines and spirits may help to communicate social status, as more generally with other forms of specialized knowledge. The forceful assertions of a connoisseur can be hard to falsify and likely serve a social function independent of knowledge of fermentations per se. As a resident of Northern California living close to the Napa and Sonoma vineyards, I am well acquainted with the pretensions and grand posturing of some aficionados. Blind wine tastings amply demonstrate that the comparative assessment of quality can be somewhat fanciful. In any event, it is clear that we are attached not just to alcohol alone, but also to an associated chemical diversity of fermentation products and flavors.

Alcohol also influences our physiology of taste via its association with food. Drink is well known to enhance the pleasure of eating, but a further and important behavioral effect is its tendency to increase overall meal size. For example, we often consume alcohol immediately prior to (as well as during) a meal. This behavior has been crystallized linguistically by the term "aperitif," which well indicates its relevance to French cuisine. The word itself derives etymologically from the Latin for "opening," suggesting both the formal initiation of a meal as well as its physical ingestion. Importantly, the consumption of alcohol with food helps us both appreciate cuisine and eat more. Although this effect may be self-evident given personal experience, it has also been studied by behavioral psychologists in a fairly systematic way. In controlled experimental settings, identical lunches were provided to fifty-two individuals and were preceded either by an alcohol-rich aperitif or by a control drink which contained either fat, protein, or carbohydrates of caloric value comparable to the alcohol. Study subjects were free to eat at will, and the sample size was sufficiently large so as to overcome potentially confounding effects of hunger and other factors among different individuals.

These experiments demonstrated a remarkable effect of alcohol alone on subsequent food consumption. Meal duration increased by about 17% on average following consumption of the alcoholic aperitif, and total energy intake increased by an impressive 30% compared to the non-alcohol controls. A number of other longer-term studies have shown, over timescales of months to years, that dietary inclusion of alcohol similarly increases caloric consumption relative to abstention. Not surprisingly, body weight tends to increase as well. This effect, in part, underlies the tendency of older individuals (who may also be drinking more) to increase in size as they age. These results, in aggregate, indicate a powerful effect of alcohol on human eating behavior. Lab rats also increase their consumption of low-concentration alcohol solutions when sucrose is added. This result clearly suggests a neurophysiological interaction between the two compounds which might advantageously serve to stimulate feeding.

It is tempting to suggest, of course, that such behavioral responses in humans and other animals originate in ancestral dietary associations of alcohol and fermenting fruit. What we do not have, however, are comparable behavioral data for food consumption by primates or other species under natural conditions. Does alcohol within fruit indeed stimulate wild animals to eat faster and to eat more over a longer period of time? This important question could potentially be addressed in the field using artificially constructed fruits with variable alcohol levels, but the logistical challenges (including variability in visitation rates and in individual hunger levels) are formidable. Nonetheless, such noise factors in data sets can usually be overcome with sufficiently high sample sizes. It sure would be fun to sit near feeding stations in tropical rainforests and to record birds and mammals gorging themselves on the experimental equivalent of cocktails at the bar. The concept of bingeing behavior in humans may have a similar analog in the wild, whereby sustained consumption of fermenting and transient resources is advantageous and is the target of natural selection.

Even if this stimulatory effect of alcohol on feeding is present in wild animals, however, such behavior may actually be self-limiting once the gut is filled with fruit or other fermenting substances. Ingestion of the structural carbohydrates, simple sugars, and, secondarily, the lipids and proteins that compose fruit tissue cannot, obviously, continue forever. When the physiological signal to stop eating is provided by distension and filling of the stomach, alcohol consumption must obviously stop as well. Dietary exposure to low concentrations of fruit-derived alcohol thus has an internal limit under natural circumstances. Humans, by

DESSERTS

CARROT CAKE 7
candied nut crumble | crispy carrot | sweetened mascarpone | caramel carrot reduction | vanilla ice cream

S'MORE 7
house made marshmallow | brownie | coffee gelato | chocolate paint | graham cracker crumble

CHOCOLATE LAVA CAKE 8
mint gelee | bing cherry | raspberry sorbet | vanilla ice cream

BUTTERMILK CHEESE CAKE IN GLASS 10

DESSERT WINES

MADEIRA | 2001 Barbieto Boal   g 10

MADEIRA | NV Rare Wine Historic Series | Charleston Sercial   g 9

MADEIRA | NV Miles 5 year   g 9

PORT | Quinta Infantado 10 year Tawny   g 8

PORT | Quinta Infantado Ruby   g 6

2007 Owen Roe LATE HARVEST SEMILLON | "Parting Glass"   g 12   b 45

2008 Adelsheim PINOT NOIR "Deglace Ice Wine" | Oregon   g 13   b 65

Figure 5. Menu with food and alcohol listings (obtained in 2011 from a restaurant featuring "innovative" cuisine in Madison, Wisconsin).

contrast, can readily circumvent this outcome through consumption of high-concentration liquid ethanol. Certainly, one effect of drinking during meals is a slowed absorption of alcohol into the bloodstream, thereby spacing out and lengthening the duration of perceived reward. But because the alcohol concentrations of modern drinks are so much higher than those occurring naturally within ripe fruit (chapter 2), much greater blood-alcohol levels can be reached prior to dietary satiation. The possibility for positive feedback, that is, further drinking stimulated by both alcohol and food consumption, then arises. Most of the time we saturate during this process at reasonably low blood-alcohol levels. However, things can go badly wrong if we don't eat at all while drinking, providing high levels of psychoactive reward but no physiological endpoint that would otherwise be associated with a full gut. Such unnatural exposure to alcohol, as with other addictive drugs, may thus stimulate pre-existing sensory biases that once usefully indicated reward. Now, however, this exposure can also result in positive and indefinite feedback that elicits dangerous levels of consumption. The challenge, then, to evolutionary biologists and neurobiologists alike, is to identify the circumstances and behavioral conditions under which such biases might have evolved.

Let us conclude this chapter with a modern dessert menu clearly suggesting the links between food and alcohol (see figure 5). The desserts, all loaded up with carbohydrates and fats, are matched against an equally impressive array of alcoholic drinks. These are classified in general terms (i.e., madeira, port, wine) but are further characterized by a much more specialized terminology, some of which will be unfamiliar to anyone other than the aficionado of sweet wines. Critically, the price per glass of these sugar-laden drinks exceeds that of any individual dessert, although much of the former consists of water, and what we are really paying for are the cultural connotations of fine fermentation and aging skills. Best of all, we can enjoy the sugar-laden desserts and sweet wines together, with the alcohol molecule as cheerful mediator linking the two in a suitable end to a rich meal. Ingestion of other drinks has presumably preceded this pleasant ending, but the parallel presentation of sweetened alcohol with high-calorie desserts nicely demonstrates ancestral association as well as ongoing significance for our modern diet. Tragically, our minds can also be misled by such positive experiences into more extreme patterns of consumption and abuse.

CHAPTER SIX

Alcoholics Aren't Anonymous

Alcoholism is a widespread, long-term, and highly damaging disease. Millions of people worldwide suffer directly from its consequences, and the indirect effects on friends, family and, in many cases, unrelated individuals are equally tragic. Although many people drink regularly with no adverse consequences, a substantial fraction will experience negative effects with varying levels of impact to themselves and others. These drinkers, and indeed many of those classified clinically as alcoholics, are often well-known socially as heavy consumers of alcohol. Others drink secretly, although family members and close friends are often aware of the problem. As a consequence, most alcoholics are in fact not anonymous. Their excessive consumption of booze is often public knowledge, as is the associated damage to self and others. What factors could possibly underlie this disease and such persistent and ultimately self-destructive behavior?

THE BURDEN OF ALCOHOL

To begin with, what exactly is alcoholism? Unfortunately, its definition has been notoriously unstable through the centuries and into recent decades. Identification of the disease also can vary among different clinical practitioners, cultural groups, and countries. In the United States, psychological disorders are routinely classified using the intermittently updated *Diagnostic and Statistical Manual* (*DSM*). Even a brief look at recent (*DSM-IV*) and earlier diagnostic criteria for alcoholism, however, is revealing. Abuse is differentiated from dependence, with the former referring primarily to the immediately negative effects of consumption, and the latter indicating long-term physiological addiction with an increasing tolerance to the drug. Sufferers of either dependence or abuse can potentially be characterized using multiple criteria, only a subset of which are necessary to establish a diagnosis. None of these refer to quantitative measures of either behavioral or physiological performance relative to alcohol ingestion. Instead, they rely on qualitative assessment of the impacts of excessive drinking on the individual. Some of the adverse consequences of drinking must be evaluated relative to the patient's social environment, making diagnosis by the clinician even more difficult.

Moreover, none of the clinically diagnostic measures refer to absolute volumes or rates of alcohol intake. Some individuals routinely consume large amounts with no hint of addiction or negative impact, whereas others suffer badly at much lower levels of intake. Withdrawal symptoms and increasing rates of drinking can be straightforward indicators of progressive addiction to the drug, but even here the absence of direct empirical measures of impact is suggestive. Clearly, we are dealing with a complicated set of biological, behavioral, and social factors that influence the likelihood someone will drink to excess. We all know drinkers for whom extreme consumption is rare and seemingly inexplicable but also very risky when it does occur. And for certain sub-populations (e.g., the homeless), alcoholism is often the norm rather than the exception. Confounding social and personal factors come into play in many of these situations, suggesting a multifaceted and complex disease.

The advent of *DSM-5* in the spring of 2013 brought yet another perspective to the diagnosis of alcoholism and of drug addictions more generally. For alcohol, addictive responses are now characterized, perhaps more accurately, as variation along a continuous spectrum of consumption, thereby merging the concepts of dependence and abuse. This approach, however, may render the distinction between regular drinking and actual alcoholism (now termed alcohol use disorder by *DSM-5*) even harder to make. Of the eleven possible symptoms of the disease, only two or more must be present to establish diagnosis. A match with three criteria is termed moderate alcohol use disorder, whereas four or more is a serious manifestation of the disease. Overall, the number of potential diagnostic symptoms for alcoholism is about the same as in *DSM-IV*, but elimination of the dependence syndrome (as distinct from alcohol abuse) perhaps sharpens our characterization of alcoholism as a unitary disease. However, its identification will continue to be challenging. Many of the numerous behaviors that continue to be indicated as diagnostic criteria, ranging from tolerance to recurrent drinking in spite of risk or social problems, will necessarily rely on self-reported information from the patient. And no underlying physiological or biochemical data will underpin the diagnosis of alcohol use disorder.

Nonetheless, clinicians can recognize most forms of alcoholism when they see it. Somewhere between 10 and 20% of regular drinkers will experience long-term adverse consequences. Of these, about two-thirds are men. For women, rates of alcoholism tend to be somewhat lower, but drinking while pregnant incurs even higher risks. Fetal alcohol syndrome is a particularly devastating disease and occurs in the United States, together with other alcohol-related neurodevelopmental disorders, at a disturbingly high frequency of about 1%. In both men and women, binge drinking is also an increasingly common problem, with a high likelihood of both acute alcohol poisoning and secondary contributions to injury and death. Such negative outcomes in the short-term are exacerbated by greater tolerance over the long run, which forces progressively higher levels of consumption. The social impact of drinking on relatives, friends, and occasionally strangers can be devastating.

At economic levels, it has been estimated that the annual cost of alcohol-related disease in the United States is on the order of several hundred billion dollars, an amount comparable to the overall health costs associated with cigarette smoking. If we then look at the global health burden (as quantified by the World Health Organization), alcohol-related disorders also turn up high up on the list. In terms of the number of human-years lost to disability and disease, alcohol use is the third-largest risk factor worldwide (after underweight childhood and unsafe sex) and poses health burdens exceeding those of unsafe water, high blood pressure, tobacco use, and overweight status. Alcoholism clearly has the potential to destroy lives, and it now affects more than one hundred million people worldwide. Given the levels of personal and societal devastation wrought by addiction to alcohol, it is remarkable that our ability even to characterize the disease using quantitative criteria remains so rudimentary.

One of the most significant impacts of alcohol consumption occurs via its interaction with motor vehicles. Although permissible limits to blood-alcohol levels are carefully legislated (with a value of 0.08% indicating drunk driving in most of the United States), the actual monitoring and enforcement of such laws are much more difficult. One of the most challenging research questions in this field is how to determine the typical alcohol load on the highways, that is, what fraction of drivers at any moment are legally intoxicated, and how does this vary with time of day and through the week? Such information is very hard to obtain absent random screening of very large numbers of motorists. Instead, statistical extrapolations must be carried out using fairly small data sets obtained from intermittent roadside checkpoints. The numbers are nonetheless revealing, if not actually frightening. Approximately one-third of road deaths in the United States, for example, are alcohol associated, and the fraction of sub-lethal but still injurious accidents associated with alcohol is similarly high. Rates of alcohol impairment are about four times higher at night than during the day and are doubled during the weekends relative to during the week.

Drivers with blood-alcohol content equal to or greater than the legal limit of 0.08% are responsible for about two-thirds of all alcohol-related road fatalities. And about one-quarter of drivers involved in fatal crashes have non-zero blood-alcohol levels, albeit ones lower than the legal limit. Clearly, the alcohol load on the roadways is dangerously high in spite of considerable policing and public policy efforts directed towards minimization of use. I certainly try to avoid driving late at night whenever possible, particularly on the weekends, when many of my fellow citizens are apparently busy boozing it up and getting behind the wheel.

One important point about such data on drinking and driving concerns the technical difficulties of accurately measuring blood-alcohol levels under field conditions. Barring an invasively drawn blood sample and subsequent analysis using gas chromatography or other methods of analytical chemistry, the vast majority of such assessments by police rely on Breathalyzers. These are small portable devices which sample our exhaled breath, oxidize the organic compounds therein, and then indirectly derive an estimate of alcohol concentration within the blood. Needless to say, a number of assumptions are embedded within this calculation. At the legal level, such ambiguities are readily exploited by the hired defenders of those accused of drunk driving. Note, for example, that the practical manual entitled *Drunk Driving Defense* is now in its seventh edition and sells online for several hundred U.S. dollars. Alcohol vapor in our breath does not necessarily correlate well with the concentration circulating in the blood and potentially varies with gender, ethnic background, body size, and the cumulatively elapsed time since starting to drink. Breathalyzer estimates necessarily ignore all of these factors and are apparently calibrated to the average human, whoever she or he may be. Intrinsic variation in estimated blood-alcohol levels, in other words, is potentially high and derives from a number of different biological factors. The aforementioned incidence rates of drunken driving should therefore be viewed somewhat conservatively.

Moreover, indirect estimates of alcohol content in the blood do not necessarily indicate the actual extent of physiological and behavioral impairment, which may vary tremendously among individuals. For some people, even one drink can slow down response times and many of the other kinds of rapid judgments that we must routinely make when driving. In many cases, the legal limits for blood-alcohol concentration are much higher than those associated with partial impairment of driving ability. Humans obviously did not evolve while driving automobiles, and the associated sensorimotor demands can sometimes be challenging even in the absence of alcohol. By slowing reaction times, impairing decision-making, and reducing neuromuscular coordination, alcohol introduces a number of potentially lethal consequences into what is already a complicated task. When driving while inebriated, we knowingly risk injury and death to ourselves and to others. This fact clearly does not deter a substantial fraction of the population from occasionally engaging in the behavior, and is powerful testimony to the overriding attractiveness of alcohol. Part of the problem simply lies in the sensory overload provided by the large amounts of the drug available today, relative to that typical of our evolutionary past. We blithely drink beyond our natural physiological limits because of the artificially enhanced and short-term psychoactive effects of alcohol. Somehow, we manage to ignore the warning messages that our brain, and sometimes the people around us, provide. This is a lethal outcome when combined with the various vehicular technologies that have turned up only in the last one hundred years or so.

Our inability to understand the basic mechanisms underpinning alcoholism is similarly mirrored in the wide range of medical and psychological approaches that have been used to treat the disease. Historically, treatments as diverse as frontal lobotomy, religious indoctrination, and even carbon dioxide inhalation have been implemented, all in the absence of data-based science that might justify such methods. That medical practitioners once resorted to such measures abundantly demonstrates the desperation of all parties involved. Today, long-term relapse rates for the sufferers of alcoholism are famously high (on the order of 90%) and have remained essentially unchanged over decades. This is not to say that some treatments aren't effective, however, with short-term success rates (on the scale of months to years) as high as 35%, depending on the approach that is used. Puzzlingly, it is also clear that spontaneous remission rates for alcoholism may be of comparable magnitude. Our predictive abilities for the likelihood of successful treatment of the disease remain an active and very important area of research.

Multiple medications are currently employed to try to treat alcoholism, of which one of the most effective is disulfiram (known commercially as Antabuse). This drug impedes the conversion of acetaldehyde in the intermediate metabolic processing of alcohol and thus causes the former to build up in the body, with its associated toxicological effects (see chapter 3 and figure 1). Interestingly, this is the same physiological outcome associated with the naturally occurring but slow-acting version of the enzyme acetaldehyde dehydrogenase (ALDH) in East Asian populations. For these groups, excessive drinking is similarly curtailed because of the toxic effects of acetaldehyde accumulation (see below). Disulfiram, via known molecular pathways associated with the metabolism of alcohol, thus reduces the likelihood of excessive consumption and deters the emergence of alcoholism. Why then is this not the wonder drug to cure the disease? Alcoholics can simply refuse to comply with treatment by not taking the prescribed drug, or in some cases will tolerate low levels of internal discomfort for the sustained buzz of drinking.

Other drugs used to treat alcoholism have more nuanced and complicated effects on brain chemistry, instead of influencing directly the physiology of alcohol metabolism. Because their underlying mechanisms of action are unclear relative to the organic basis of the disease, these pharmacological agents are much less likely to have positive effects. Instead, they must be prescribed in almost a trial-and-error method in the hopes of success, and individual sufferers vary dramatically in their responses. Only two such drugs are currently approved by the U.S. Food and Drug Administration to treat alcoholism, namely acamprosate and naltrexone. Both tend to reduce the psychological craving to drink, but only for about 14% of treated alcoholics. This is clearly a hit-or-miss approach to medicine absent basic understanding of the neurophysiology and chemistry of the desire to consume alcohol. But the good news is that they do work for the lucky few, albeit for reasons unknown. At least four other drugs unapproved for clinical use are similarly being investigated to manage alcoholism. The testing of such drugs represents experimental medicine at its finest but also indicates the conceptual limits to this approach. Neurochemistry of the mammalian brain is famously complex, as we might expect from a structure comprising billions of neurons, each with thousands of connections. Predicting the action of individual drugs on a complex behavioral disorder such as alcoholism lies well beyond current understanding of brain function.

In contrast to such pharmacological interventions, counseling and psychotherapy are also routinely used to treat alcoholism. Given that much of the disease is defined relative to its effects on the alcoholic's social and personal context, therapeutic treatment can be beneficially targeted towards these external factors, particularly to try to mitigate any associated risks of drinking. As a consequence, the prohibition of alcohol intake is not necessarily indicated as a routine part of such treatment. Indeed, if regular dietary exposure to alcohol is an engrained feature of our evolutionary past, then abstention will be very difficult to enforce. And as sufferers of alcoholism and their family members know all too well, seemingly irrational behaviors and rejection of therapy are often the norm for heavy drinkers. The best planned course of physician-advised recovery can be overturned in a split-second decision to resume drinking. Absent understanding of the biological mechanisms responsible for addiction to alcohol, it is not surprising that medical treatments have been and remain fairly unsuccessful. But given the themes of comparative biology as developed throughout this book, an evolutionary perspective can certainly be informative. Behaviors only evolve when selection acts on inherited traits, so it is helpful to look at genetic associations of the disease.

IT'S IN THE BLOOD

Alcoholism has long been known to run in families. This observation alone, however, does not necessarily establish hereditary effects, as common environmental factors within a family may also elicit a tendency towards excessive drinking. To unequivocally establish genetic underpinnings, it is instead necessary either to manipulate the genome directly and look at the phenotypic outcome (which is easily done in fruit flies or mice but is much less feasible in humans) or to carry out multigenerational studies of families. The best way of doing this for complex behavioral traits in humans is to study split twins, fraternal or identical twins who were separated shortly after birth and who then grew up in different environments. This approach, utilizing the known extent of shared genetic background, enables the fractional variation in adult behavior that is attributable to environmental effects alone to be quantified. Such studies suggest values of heritability for alcoholism of 0.2 to 0.6, whereby a value of zero indicates an entirely environmentally determined outcome, and a value of one indicates a perfectly inherited trait. The numbers obtained to date suggest that genetic and environmental effects are of comparable significance with respect to the likelihood of developing alcoholism.

But what exactly is being inherited here that predisposes one to the disease? And what kinds of environments tend to elicit, perhaps in combination with a particular genetic background, our addictive responses to alcohol? Here we enter a complex domain of biology, wherein networks of many different gene products interact with their cellular surroundings and with one another. From moment to moment, the massive numbers of cells that compose our bodies (thought to be on the order of hundreds of trillions per body) also receive signals from their neighbors, from the food we have ingested, from our social interactions with those around us, and from the physical environment in which we live. The DNA within our cells encodes about 25,000 genes in total, hundreds if not thousands of which potentially influence the tendency to drink to excess. Identifying and deciphering the numerous ways that such genes might contribute to the disease, as well as how these effects might change ontogenetically (i.e., through embryonic development, infancy, childhood, and into adulthood), and in response to different social and physiological conditions, has thus far been an insurmountable problem for addiction professionals. Epigenetic effects, which involve non-heritable changes to gene expression and regulation that nonetheless persist across multiple generations, further confound the issue.

Furthermore, most behavioral disorders such as alcoholism are influenced by the action of not just one gene. Unlike relatively simple diseases such as sickle-cell anemia (which derives from a single mutation in the gene encoding the hemoglobin molecule), most human phenotypes typically reflect the contributions of many genes. Some of these may interact with specific features of the surrounding environment to further influence the likelihood that a particular disease will appear. Because such effects have not been elucidated for any of the genes potentially associated with alcoholism, our ability to diagnose and treat the disease at such a reductionistic level is essentially nonexistent. Addiction to alcohol also involves diverse responses of multiple organ systems in the body. In the brain, the disease appears to associate with long-term changes in different kinds of neuronal connections (i.e., the synapses) and in other aspects of cellular physiology. Alcohol is a wide-acting drug with effects on many different constituents of the central nervous system and on other regions of the body (chapter 5). The molecular effects of prolonged exposure are thus multifaceted and influence many different features of our physiology and neurochemistry. This outcome makes it very difficult to associate particular genetic backgrounds and inherited signaling pathways of the brain with the clinical phenomenon we term addiction.

Given the large number of sufferers of alcoholism, however, it has been possible to use statistical approaches to identify certain behavioral correlates of the disease. Many alcoholics tend to start drinking in adolescence and early adulthood, and exhibit reduced impulse control in other aspects of their lives as well. Stress is another factor frequently implicated in the emergence of alcoholism in humans, and this response may also characterize other animal species. For example, mice that have been selectively bred to have a dysfunctional hormonal pathway involved in regulating their stress response drink markedly more alcohol (at concentrations up to 8%) when socially stressed. Similarly, work with one species of Old World monkey has shown that voluntary drinking of alcohol is more likely following social separation, in what is otherwise a highly gregarious primate species. Genetic variation in the response to stress, and in other personality traits relating to self-control, may thus indirectly influence chronic responses to alcohol. But these emergent features of behavior must also involve other molecular signaling pathways and their associated gene-encoded proteins within our brains. As with various rodent studies (as discussed below), considerable effort has been devoted to identifying candidate genes potentially responsible for alcoholism in humans. These studies have been largely inconclusive, however, and have now been superseded by a more sophisticated approach (termed genome-wide association) which evaluates single-nucleotide variation in the DNA of thousands of individuals. To date, this approach has failed to confirm earlier identifications of putative candidate genes for alcoholism. It is clear that unambiguous association of the disease with particular genetic markers remains an important future goal for such studies. Experimental studies which knock out particular genes in animal models, so as to test directly their functional consequences, are perhaps the best means of assessing causation relative to addictive responses.

We must keep in mind, however, that it is difficult to uniquely identify the phenotype we term alcoholism. Our concept of this disease, as manifested in modern humans, likely conflates a diverse assemblage of differing behavioral and physiological responses to alcohol. These behaviors can occur on widely varying timescales, from short-term attraction to the odor and taste of the molecule to the tendency to drink heavily over many decades. Are these responses necessarily the same, and are they going to be linked genetically? Or is our operational characterization of the disease simply inadequate to incorporate the varied biological features of the behavior, however tempting it may be to uniquely pathologize the syndrome? For example, addictive responses derived from the reinforcing effects of repeated high-level drinking may act independently of feeding behaviors associated with ancestral dietary exposure to low-level alcohol. Addictions more generally, be they in the domain of drugs, foods, gambling, or other arenas of human life, ultimately must derive from reduced self-control. The neural structures responsible for limiting response in the face of reward are controversial, although the limbic system and prefrontal cortex of the brain are routinely implicated. It is clear that the associated dopaminergic reward pathways play a major role in addictive behaviors and are activated by many kinds of drugs consumed by humans, including alcohol. Inherited variability in these brain regions may well underlie some aspects of drug use, but the extent to which both low-level alcohol consumption as well as its extremes reflect such genetic influences is unknown.

A further and fascinating hint as to the origins of alcoholism lies in its association with a preference for sweets. Although largely anecdotal, this outcome has nonetheless received some systematic attention. Self-reported pleasurable responses to sugar solutions of different concentrations tend to be higher for those with family histories of alcoholism. And studies with rodents and monkeys have also suggested a genetically based correlation between the tendency to drink alcohol and the preference for sugar. For well-grounded evolutionary reasons, this association can be no accident. Both alcohol and sugar are found primarily within nutritious fruit pulp, and the former compound derives uniquely from the fermentation of the latter (chapter 2). Similar reward pathways (involving both the dopaminergic mesolimbic system and the opioid system) act to regulate pleasurable responses to these molecules. Metabolic responses to excessive consumption, occurring both in the liver and in hormonal terms of insulin regulation, are also comparable for both substances. Moreover, ingestion of sugars while drinking (think of wine, beer, or any mixed drink for that matter, all of which contain substantial carbohydrates) increases significantly the rate of alcohol metabolism via synergistic effects on liver enzymes. All told, these outcomes indicate a broad congruence for sugar and alcohol in terms of biochemical processing and metabolic habituation, as well as in our responses to high levels of exposure. It is tempting to suggest that addiction to either compound simply represents amplified but ultimately maladaptive outcomes, given that the amounts available in today's world are abnormally high relative to those obtained through evolutionary time. We would also predict, however, a continuous spectrum of behavioral responses given the nutritional and physiological benefits associated with more natural low-level consumption.

Might, in fact, all drug addictions have a similar origin? Many psychoactive compounds used recreationally by humans over the centuries are natural substances produced by different kinds of plants. In one critical aspect, however, alcohol differs fundamentally from other types of recreational drugs (including nicotine, caffeine, opiates, and many other alkaloids). Via toxic and in some cases psychoactive effects on the nervous system of herbivores such as caterpillars and grazing mammals, these compounds act to deter the consumption of leaves and seeds. Such specialized chemicals, however, are taxonomically fairly specific with respect to the kinds of plants that produce them, and they tend to be sparsely distributed within natural habitats. High and sustained levels of consumption by primates and other animals would correspondingly be unlikely. Alcohol, by contrast, is widely distributed within many different kinds of ripe fruit in tropical environments and features in the diet of most primates as well as those of other fruit eaters. What we now see as addiction may simply reflect positive reinforcement via neural reward pathways that were once highly useful given daily low-level exposure. Today, these pathways can be overwhelmed by the higher blood-alcohol concentrations attained courtesy of controlled fermentation and distillation. Other recreational drugs extracted by humans from plants, as well as those synthesized artificially, may simply be piggybacking chemically onto these pre-existing reward circuits. Addiction biology more generally might thus be derived from the natural occurrence of fermentation and from those forces of natural selection that have acted on the behavioral responses of primates to alcohol.

Our consumption rates today, by contrast, are ultimately demand- and not supply-limited. Unfortunately, the high levels of alcohol we can potentially indulge in may elicit adverse responses from our genome. Other chemical compounds characterized by similar U-shaped dosage-response curves (i.e., hormetic substances) also yield bad effects under abnormally high exposure. Sugars and animal fats are two of the most obvious examples, which along with alcohol can be mass-produced at very low cost. For alcohol, it's not that routine low-level consumption is necessarily bad (with much evidence suggesting it is actually beneficial), but simply that today's amounts and concentrations can easily facilitate unnatural levels of exposure. Nonetheless, it is unclear why most individuals can well tolerate some drinking and process alcohol normally, whereas a minority drift into syndromes of addiction and disease. Underlying patterns of genetic variation must be part of this puzzle. And one of the most interesting observations in alcohol research is that contemporary human populations differ considerably in their responses to the molecule.

RED FACES AMONG THE CHINESE

Some people can drink a lot with no obvious behavioral effects, whereas others are highly sensitive to even very small amounts of alcohol. In East Asia, this latter outcome is actually the norm rather than the exception. Although Asia is a cultural construct that corresponds to no particular set of geographical boundaries, it is clear that many people within the contemporary political regions of Japan, Korea, and China simply can't drink at all. Even small amounts of alcohol induce low-level negative effects such as flushed faces, increased perspiration, and general unease, along with more substantial adverse outcomes, including elevated heart rates, wooziness, nausea, and passing out. My own in-laws and other relatives in Hebei are extremely wary around booze, and when they do drink will take only the smallest sips well spaced over time so as to limit dosage. Anything in excess of this (i.e., what a typical north European would consume in one gulp) is enough to make them leave the table.

This is a remarkable reaction to alcohol relative to that of most Europeans, North Americans, and many others worldwide. Moreover, this effect can be directly attributed to genetic differences among different human groups. In these Asian populations, adverse physiological responses derive directly from the expression of particular enzymes that metabolically degrade alcohol. As in fruit flies, the human genes that encode for both ADH and ALDH (see figure 1) are variable. First, the ADH enzyme that acts initially to transform and metabolize alcohol molecules has one form that occurs at a very high frequency in East Asia relative to other parts of the world. This form of the enzyme acts relatively quickly, enhancing buildup of the ensuing metabolic product, acetaldehyde. Second, the ALDH enzyme that then helps to degrade this intermediate molecule is very slow acting in many of the same human populations. The net effect is rapid degradation of alcohol, but simultaneously an enhanced buildup of acetaldehyde. This molecule can be toxic at even very low concentrations, resulting in a broad suite of unfortunate physiological reactions for the drinker in question. In aggregate, these effects have been informally termed the "red face" syndrome by the medical profession and are a reliable marker for acetaldehyde accumulation while drinking. The ADH and ALDH genes are on separate chromosomes and are thus not in genetic linkage, so the fact that the allelic forms present in East Asia act together to yield this pronounced physiological effect is of considerable evolutionary significance.

Interestingly, alleles for the slow-acting ALDH enzyme also occur at high frequencies in many indigenous South Americans, consistent with the historical derivation of the latter group from migrations out of northeastern Asia. For the indigenous peoples of North America, however, the situation is much more complicated. In part, the underlying genetic heritage seems to be more complex than is the cultural expression of tribal identity. For one thing, there has been considerable genetic mixing between indigenous groups and colonizing European peoples over the last five hundred years. Neither the ADH nor the ALDH in those North American indigenous groups sampled to date differs substantially from those characteristic of northern Europeans. And their physiological ability to clear consumed alcohol is also very similar, in contrast to anecdotal accounts of a greater susceptibility of some native peoples to "firewater." It is important to emphasize here the tremendous genetic diversity that characterizes both North and South American indigenous groups, most of whom have not been systematically studied with respect to the capacity to metabolize booze.

Also, cultural differences and sustained exposure to alcohol in more recent times will influence both physiological and behavioral responses to the molecule. The rates of alcoholism among indigenous North and South Americans tend to be somewhat similar, in spite of the much higher occurrence of slow-acting ALDH in the latter group. It is also clear that a number of different social and economic factors impinge on drinking outcomes for these indigenous groups, not least of which is poverty. And given the somewhat labile definition of alcoholism, it is important to be skeptical about such broad cross-cultural comparisons. Some native peoples of the New World also have highly divergent diets relative to those of their ancestors in Asia. Inuit and other indigenous groups of northern Canada and Siberia, for example, have historically consumed virtually no carbohydrate except for small summer fruits such as blueberries and hackberries (which are unlikely to contain much alcohol; see chapter 2). These northern peoples, interestingly enough, have been reported to metabolize alcohol more slowly than do indigenous groups from more equatorial latitudes, although the underlying genetic factors have never been investigated.

Given that East Asian populations are characterized by very different forms of enzymes that influence alcohol metabolism, an obvious question concerns the historical origins of this geographical pattern. How old are the alleles that encode these particular enzymes? Are they about 10,000 years old, placing them near the origins of agriculture? Or are they on the order of millions of years old? And what selective factors might have acted so dramatically to increase these gene frequencies in local populations relative to other parts of the world? To answer these questions, it is first necessary to sample extensively within East Asia to obtain a more finely resolved understanding of the local distribution of the alleles. Such work has been carried out thus far only for the fast-acting ADH allele in mainland China, but the results are nonetheless remarkable. As one moves from the eastern coast of China inland (i.e., starting in the vicinity of Shanghai and spreading concentrically outwards), the high frequencies of this allele slowly drop until, at a distance of more than 2,000 kilometers from the starting point, the incidence is approximately equal to the low background level found in the rest of the world. The frequencies of the allele in Korea and Japan are also high and are comparable to those on the east coast of China. This result is consistent with hypothesized population origins for these two regions within the East Asian mainland.

Overall, the pattern of ADH distribution in East Asia represents a remarkable gradient (or cline, as it is known to geneticists), and it is highly unlikely to have arisen by drift or random effects alone. Using extensive geographical sampling, modern methods of DNA sequencing, and sophisticated phylogenetic methods, the evolutionary history of this fast-acting allele has been reconstructed. With such approaches, the emergence time of this particular ADH allele in China has been bracketed statistically to somewhere between 10,000 and 7,000 years ago. Perhaps not coincidentally, this timing correlates well with the origins and spread of rice cultivation over the same geographical area. Archaeological evidence documenting the domestication of rice is well studied and indicates a spread westward and southward from central China starting about 12,000 BCE. Interestingly, various ethnic groups on the periphery of historical Han Chinese civilization (e.g., Manchurians, Mongolians, and Tibetans) are more alcohol-tolerant and drink various kinds of fermented beverages (e.g., *chaang,* a kind of barley beer in Tibet, and *kumis,* the fermented mare's milk of Mongolia). Evidence to date suggests that a slower-acting ADH allele is dominant in these populations, in contrast to the faster rates of alcohol metabolism characteristic of the rice-based cultures in eastern and central China.

These studies thus suggest that the genetic tendency for acetaldehyde accumulation is a fairly recent outcome in East Asia, and that it turns up along with incipient agriculture in the same region. An inevitable question then concerns those evolutionary forces that have maintained alleles for fast-acting ADH at such high frequencies but also within a limited region of the world. Similar effects have presumably selected as well for the slow-acting ALDH alleles in these populations. Here we have much less evidence, but there are a number of intriguing evolutionary possibilities. Selection may have acted directly to deter consumption of alcohol via toxic and aversive effects associated with accumulation of the metabolic intermediate of acetaldehyde. There may well be advantages to avoiding both the direct consequences of inebriation and the potentially negative long-term effects of high exposure. As we will see shortly, modern rates of alcoholism are fairly low in East Asia relative to many other regions of the world, although what the selective context for this outcome would have been 10,000 years ago can only be guessed at. If these East Asian alleles do indeed protect against alcohol consumption, then the beneficial effects of low-level alcohol consumption (as discussed in chapter 3) might not pertain for those individuals who, for genetically-based reasons, either never or only rarely consume alcohol. Unfortunately, the appropriate epidemiological studies have only been carried out in North America and western Europe and have not explicitly teased out the hormetic consequences of variation in these alcohol-related genes.

Another possibility concerns the presence of an additional biological participant in the balance of selective forces acting on the ability to metabolize alcohol. For example, the gene responsible for sickle-cell anemia in humans persists mostly in sub-Saharan African populations because heterozygote carriers of the relevant allele have a higher resistance to malaria. The time-averaged balance of selection between these different costs and benefits thus maintains the sickle-cell allele at non-trivial frequencies. Comparable mutations in hemoglobin genes can also be found in Southeast Asia, where malaria is endemic. The signature of selection imposed by malarial parasites over hundreds of thousands of years can thus be read within the genome of present-day human populations. For many East Asians, the inability to tolerate alcohol may similarly reflect the action of additional biological players. One possibility is that fungal poisons associated with the storage of rice may be detoxified by higher acetaldehyde concentrations in the human body. Many different kinds of fungal pathogens can attack rice crops and stored products, however, yielding a wide array of potential toxins which may not necessarily be sensitive to physiological concentrations of acetaldehyde. More specific is the idea that resistance to the viral disease hepatitis B is greater in acetaldehyde accumulators. Hepatitis B is endemic in eastern China and correlates well with the geographical distribution of the slow-acting ALDH allele. It is also known to interact with long-term rates of alcohol consumption to influence the likelihood of developing liver cirrhosis and cancer, although the specific molecular mechanisms involved in this process are unknown. One way to test this hypothesis would be to evaluate whether carriers of the slow-acting ALDH allele also have lower rates of mortality from hepatitis B and associated liver diseases. Moreover, acetaldehyde has been implicated in certain cancers of the upper digestive tract, the occurrence of which can obviously be minimized if alcohol is avoided altogether.

Diagnostic problems notwithstanding, the incidence of alcoholism also tends to correlate with aforementioned geographical patterns of variability in alcohol metabolism. This should come as no surprise. Those individuals who, for reasons of genetic background, cannot tolerate alcohol simply tend to avoid it altogether. By definition, people can't become alcoholics if they never drink alcohol. From a large-scale geographical perspective, it has also been historically noted that rates of alcoholism among East Asians tend to be much lower than those in European and most North American populations. Presumably this result derives from the deterrent effects of acetaldehyde accumulation given slow-acting ALDH enzymes within the East Asian populations. Variation in addiction to alcohol, however, could derive from a number of other factors, of which genetics might be only one. Cultural differences in the assessment of the disease, as well as variation in drinking practices and outcomes, render such broad-scale intercultural comparisons only suggestive. What is really required is a tighter linkage over much smaller geographical scales of the tendency to drink excessively, together with description of the corresponding physiology of alcohol degradation.

Fortunately, such information has been obtained for multiple East Asian populations over the last decade or so. Independent studies conducted with large groups of Taiwanese, Japanese, and Korean alcoholics found a remarkable association between the propensity to drink and variability in the metabolic pathways that degrade both alcohol and acetaldehyde. Relative to control populations, those individuals characterized as alcoholics (using local medical criteria) were as much as ten times likelier to have a slow-acting ADH enzyme (thus reducing acetaldehyde buildup), as well as faster-acting ALDH enzymes (which would act to degrade this metabolic toxin). In other words, these alcoholics were genetically much more similar in this regard to western Europeans than to their fellow East Asians. As a consequence, no deterrent effect of acetaldehyde accumulation pertains, and the individuals concerned are physiologically free to indulge at will. This doesn't necessarily explain why they might ultimately become addicted (however this concept may be defined), but it certainly underpins much of their long-term alcohol consumption. Environmental influences presumably play a role here as well, but the explanatory power of variation in these two genes is nonetheless striking. Moreover, some rare variants of ADH may also influence the susceptibility of non-Asians to become alcoholics.

In addition to ADH and ALDH, genes for many other enzymes involved in the human diet have undergone selection in relatively recent times, that is, since the origins of agriculture. The best studied of these is the trait of lactose intolerance, which refers to the inability of many adults worldwide to metabolize the predominant sugar in milk. The major exceptions to this trend lie among northern Europeans and some African populations who, for historical reasons deriving from cattle domestication some 10,000 years ago, retained the ability to digest milk via expression in adults of the lactase enzyme. In these groups, dairying and the associated spread of farming provided a net caloric gain which apparently was the target of positive selection, promoting retention from infancy of the ability to digest milk. Other aspects of the modern human diet have similarly been modified by natural selection over fairly short time periods, including some genes involved in taste reception and the enhanced expression of amylase (an enzyme that degrades starch). Rapid evolution of human nutritional physiology is thus well documented and highly feasible given appropriate evolutionary circumstances. For alcohol, however, the selective forces on those alleles influencing its metabolic degradation are an unknown, albeit tantalizing, source of information about our long-term genetically based responses to exposure.

Substantial costs as well as benefits derive from the consumption of alcohol, but our understanding of the tendency to drink to excess remains poor at best. Treatments for alcoholism are famously unsuccessful, and the carnage from drunk driving on the roads continues unabated. The only known protection against the disease is the genetically based expression of particular ADH and ALDH enzymes involved in the metabolism of alcohol, mostly confined to East Asia. Unfortunately, individuals have no control over their inherited genetic background, and the likelihood of developing alcoholism can't really be predicted accurately for most people. We also can't carry out on humans the kinds of multigenerational manipulations that are now routine for various animal systems in biomedical research. Use of the latter, including the fruit flies discussed in chapter 3, has been informative in identifying common molecular underpinnings to inebriation but has been much less useful in broader behavioral interpretations of the disease. Nowhere are such limits more apparent than in studies of rodent and primate responses to alcohol.

THE CHIMP PREFERS MARTINIS

One powerful approach for studying medical problems in humans is to develop comparable models of disease in other species. The major advantage here is the possibility of direct experimentation and manipulation of those various factors, including genetic components, that influence the expression of particular pathological outcomes. In biomedical studies of alcoholism, animal models using mice, rats, monkeys, and apes have been particularly important. Rodents provide for large sample sizes and well-established physiological protocols, and the study of primates provides for some degree of evolutionary similarity to humans. The short life span of rodents also permits artificial selection for genetically based addictive behaviors, including consumption and withdrawal symptoms, over many generations. Mice and rats can be selectively bred in cages for decades, and the use of crosses among strains and modern molecular approaches (e.g., gene knockouts) allows for particular behavioral responses to be pinpointed in the genome. The fact that artificially imposed selection for the tendency to drink liquid alcohol yields rodent strains that consume progressively greater amounts well demonstrates a heritable component of this aspect of addiction. Much research has correspondingly gone into understanding the genetic, neurological, and behavioral foundations of these responses.

Nonetheless, such experiments have led to only limited insights into the basic biology of human alcoholism. The very use of the phrase "animal model" conveniently ignores the fact that humans too are animals, albeit fairly complicated ones. And naturally occurring ingestion of alcohol via a fruit-based diet more broadly characterizes many primates, including our own ancestors. Most studies with rodents and other non-human taxa explicitly attempt to simulate alcohol consumption as it occurs in modern humans, providing alcohol in a dilute, watered-down form as an adjunct to an otherwise solid diet. In the real world, of course, alcohol and nutritional substrate are inextricably bound together within the bodies of fermenting fruits. Lab experiments to date simply haven't captured the complexity of animal feeding responses to simultaneously presented alcohol and dietary calories. More broadly neglected in addiction research is any evolutionary perspective on the study species, including our own, and reconstructions of likely historical exposure to alcohol are absent.

For the rodents typically used as standard models in biomedical research (the Norwegian rat and the house mouse, and sometimes the hamster), natural exposure to alcohol has in fact been negligible. All of these species occur normally only in the temperate zone, where they would have had little or no access to naturally occurring alcohol. Instead, these species are predominantly omnivorous (but with a preference for grains) and exhibit no particular specialization on fruit. There are certainly many tropical rodents that scavenge on ripe and fermenting fruit, but for a variety of operational and historical reasons, these species have not been used in biomedical research. Lab rodents do indeed exhibit a number of different behavioral responses to dilute alcohol solutions, including escalation of drinking, increased tolerance, and withdrawal symptoms. It is not, however, surprising to learn that use of rodent models has reached important experimental limits. It has proven particularly hard to get the animals to drink alcohol at the very high rates that would correspond to drinking by human alcoholics (given appropriate correction for the very large differences in body mass). Given their absence of historical exposure to the molecule, this is a probably a pre-ordained outcome.

Furthermore, definitions of alcoholism in humans are notoriously flexible (see above) and often rely on uniquely human social and behavioral contexts that cannot, by definition, pertain to non-human mammalian species. What is termed alcoholic behavior in humans is often evaluated relative to its deleterious consequences within an individual's social environment. This factor is obviously difficult to reasonably mimic in a rodent colony. Instead, the experimental focus has been on the physiological concentration of alcohol that is reached and sustained in the blood stream, independent of long-term behavioral outcomes. However, many of the drinking patterns seen in lab animals may simply derive from their very different sensory and neural physiologies; rodents are not primates, after all. It is much more likely that the outcome characterized in rodents as elevated drinking actually conflates a number of different behavioral responses (including reactions to novel or abnormal nutritional cues and to the cage environment) that ultimately are manifested as excessive alcohol consumption. And although alcoholism in humans likely involves many genes, as well as strong environmental influences (see above), these factors may differ from those identified to date in animal models.

Nonetheless, by selecting and artificially breeding for high- and low-drinking behaviors, different strains of mice and rats have been generated that are genetically variable in their behavioral responses to alcohol. It is unclear, however, if these behaviors are physiologically analogous to similar patterns of variation in modern humans. In some of these rodent strains, candidate genes have been identified that may correlate with drinking tendency. For example, some such genes have been found that statistically associate with higher rates of alcohol consumption when rodents are given, in binary choice tests, comparable drinks with no alcohol content but equivalent energetic value. The proteins encoded by these candidate genes serve a variety of cellular functions. However, none of these have any necessary or obvious relationship with either addictive behaviors or direct metabolic responses to alcohol. The implications of these candidate genes in rodents for our understanding of human alcoholism remain unclear. And even the validity of the approach has been contested, as behavioral traits are typically influenced by many interacting genes and by environmental factors, which aren't assessed in these studies.

The use of monkeys and the great apes to simulate patterns of human drinking has also been fraught with problems. Basically, giving a large caged primate a choice among alcohol solutions of different concentrations over different time intervals isn't going to tell us much about the biological foundations of drinking behavior. A high level of individual variability, the effects of laboratory confinement, and social factors are all confounding factors in such research. Sample sizes are also necessarily limited because of the high expenses associated with maintaining primate colonies. Although such studies in the 1970s with chimpanzees demonstrated both increasing tolerance to and withdrawal symptoms from alcohol, more recent work has utilized rhesus macaques, an omnivorous Asian species of primate widely used in biomedical science. Both male sex and juvenile stress have been shown to be risk factors in this species for high levels of alcohol consumption, just as they are in modern humans (see above). A major difficulty in such work, however, derives from the operational definition used for addiction. Individual animals may vary substantially in the daily mass of alcohol consumed relative to body mass and in the fraction of daily energetic expenditure that derives from the metabolic oxidation of alcohol. Both aspects of alcohol use may change with activity patterns and with other avenues of energy expenditure, particularly in social animals. Sometimes individual test subjects have been labeled as light, medium, or heavy drinkers (based on their relative rates of consumption), but this approach ignores the natural continuum of behavioral responses to alcohol. It also anthropomorphically projects onto other animal species certain patterns of human drinking that simply may not be biologically relevant.

And for all of the fly, rodent, and primate studies of alcoholic behavior, a broader question concerns the significance of these for excessive drinking by modern humans. Alcoholism derives from a large number of interacting factors, including the action of many different genes, the particular social and environmental context of exposure to alcohol, and an individual's personal history. These factors are obviously not possible to replicate in animal systems, other than to provide superficial analogies to such broad-based factors as stress and alcohol availability. Voluntary drinking of liquid alcohol is simply not relevant to the natural biology of the rodent and primate models used in these studies. The addiction researchers working with different mammal species have, in this case, truly failed to see the (rain)forest for the trees. For larval and adult fruit flies, consumption of alcohol-laced food reasonably approximates conditions experienced in the wild, but the ecological context of exposure is obviously very different from that in humans. And those biological factors predisposing wild animals to excessive alcohol consumption may differ from those characterizing our drinking patterns today.

For example, the likelihood for humans to have driving accidents must ultimately derive from a variety of physiological and psychological factors. But could we reasonably use primates, rodents, or fruit flies to study this problem? We could potentially try to understand traffic accidents by placing monkeys behind the wheel under simulated driving conditions in the laboratory. But any attempt to correlate outcomes with particular genetic and behavioral traits of individual monkeys is likely to tell us next to nothing about why modern humans might suffer driving accidents. Much more informative would be identification in humans and, comparatively, in other animals of those more fundamental sensory and motor tasks involved in fast reactions (e.g., multitasking and visual tracking of one's surroundings). Then the underlying physiology and genetic underpinnings of such behaviors might be better understood. These factors are, of course, part of our ancestral baggage as large bipedal social primates. Similarly, the human response to alcohol must derive, at least in part, from inherited functions of our sensory and behavioral biology. At this stage, understanding the natural behavior of animals towards fermenting fruit might well be a more productive line of enquiry if we are to understand what motivates our own alcoholic behavior.

CHAPTER SEVEN

Winos in the Mist

The drunken monkey hypothesis proposes that our contemporary responses to alcohol, both positive and negative, are in part inherited from our primate ancestors. In science, as distinct from many other philosophical and cultural enterprises, the ultimate test for any given claim about reality is falsifiability. Data can be systematically collected in the real world to test the likelihood that any hypothesis posed a priori is, in fact, wrong. In this chapter, I discuss future research directions that can test some of the key ideas and predictions presented throughout this book. In the wild, what are typical fruit-alcohol concentrations? To what extent and how frequently do primates and other fruit-eating animals get exposed to alcohol? How does such natural exposure influence feeding behavior and, more generally, addictive responses in primates and other animals? And to what extent are these behaviors genetically based?

These are just some of the many experimental questions that can emerge from a comparative approach to the natural biology of alcohol. Wild chimps on booze and Breathalyzers, figuratively speaking, is the entertaining research agenda outlined here. At various points in this book, I have proposed new interpretations for those behavioral motivations common to routine alcohol consumption and to addiction. And the evolutionary perspective suggests a number of observational and experimental possibilities that might help to elucidate the relevance of alcohol exposure in nature. In addition to such data-based approaches, it is essential to appreciate the complexities associated with interpreting modern human behavior relative to historical antecedents, as well as the need for responsible drinking in today's world. But let us first lift a glass in praise of Charles Darwin, whose keen and penetrating insights continue to infuse the minds of hundreds of thousands of biologists alive today and marveling at biological design.

DARWIN IN THE GIN PALACE

Pigeon breeding, one of the many interesting cultural trends in Victorian England, fascinated Charles Darwin. The domestication of animals, as an exemplar of artificial selection, was of great interest to Darwin as it demonstrated the possibility for rapid changes in both morphological and behavioral traits. Domesticated breeds of animals, including pigeons, were also readily accessible and often had known pedigrees. Pigeon breeding in particular was a vigorous Victorian hobby cutting across a wide range of social strata. Although a wealthy patrician and landowner, Darwin belonged to several workingmen's pigeon clubs as well as to a more elite breeding association. At home at Down House, he also carried out experiments in pigeon breeding, the results of which buttressed his emerging theory of biotic evolution via the two mechanisms of natural and sexual selection. On a number of occasions, Darwin ventured into London to meet with pigeon fanciers at Borough Market near London Bridge. Such meetings were usually held in public houses (i.e., pubs), during which breeders would show off live examples from their collections. Such displays amply demonstrated the impressive range of morphological diversity that had been elicited by intense artificial selection over just a few generations.

From his letters, we know that Darwin's attendance at these meetings and during other intermittent trips to London was accompanied by claret and by "quantum suffi" (a sufficient quantity) of other kinds of wines. It is clear that throughout his life, he was not averse to moderate drinking. As an undergraduate at Cambridge University, he was a member of an eating club and sometimes drank to excess. The HMS *Beagle*, on which Darwin served as naturalist from 1831 to 1836, was well-provisioned with rum (via a spirit room belowdecks), thanks to its official demarcation as a vessel of the Royal Navy. At Down House, Darwin apparently drank only small amounts (e.g., one glass of wine a day, as well as ale), but he consumed this with great pleasure. Both brandy and wine were occasionally prescribed to him for medicinal purposes. And he once expressed horror at the possibility of drinking too much, perhaps because of a family history of alcoholism several generations past.

It is perhaps unsurprising that Darwin, given his wide-ranging interests in natural history, also commented in his voluminous writings on animal inebriation. At one point, he cited a zoological compendium by the German biologist Alfred Brehm to the effect that African baboons could be attracted using strong beer, and suggested that in this regard the taste nerves of monkeys and men must accordingly be similar. Although Darwin wrote extensively on diverse themes in human evolution, as exemplified by his two books on the subject (*The Descent of Man, and Selection in Relation to Sex,* and *The Expression of the Emotions in Man and Animals*), he paid little attention to the interesting question of diet. However, in a letter written on 11 September 1877 to W. M. Moorsom, he suggested that most monkeys would regularly consume alcohol were it available, and he referenced captive monkeys held by a publican that were regularly given alcohol so as to elicit drunkenness. Although he drew no direct connection between fruit-eating and the attraction to alcohol, the conclusion that some non-human primates exhibit innate behavioral responses to the molecule was nonetheless insightful.

In fact, evolutionary perspectives on what we eat and drink really only got going in the 1970s and 1980s. With the appearance of such works as *The Paleolithic Prescription* and numerous scientific articles, researchers attempted to place modern food choices within a deeper biological context going back millions of years. Several decades of thought-provoking work on human diet have subsequently encompassed paleontological findings on ape dentition, the foraging strategies of extant hunter-gatherer societies, and the molecular evolution of taste genes, among other themes. In parallel, escalating epidemics of such metabolic diseases as diabetes and obesity have focused attention on understanding diet as a set of behavioral choices influenced at least partially by historical factors. In most industrialized countries, and increasingly in developing countries as well, the consumption of calories has switched from being supply limited to being demand limited. Given the essentially unrestricted access to cheap meat, animal fats, and processed sugar that industrial agriculture can provide, the concept of modern food as a potentially addictive substance has to be taken seriously.

Our consumption of alcohol today, as argued throughout this book, can be viewed similarly. In preceding chapters, I have developed the theory that natural selection acted on our evolutionary forebears to associate the presence of alcohol with nutritional reward. The primary benefit from this linkage was the ability to rapidly find and then consume those sugar-rich fruits intrinsic to the primate diet. As with the pleasurable rewards associated with ingestion of fats, sugar, and animal protein, so too does exposure to alcohol psychoactively stimulate further consumption. This cycle results in a repeated reward loop that reinforces the behavior. For some individuals, such positive feedback can ultimately result in uncontrollable rates of ingestion and a cascade of long-term pathological changes. Whereas once our exposure would have been limited at the end of a meal of alcohol-containing fruit, now the higher concentrations of alcohol available in liquid form allow us to overcome this limiting mechanism. Addiction, indicating progressively higher and uncontrollable rates of consumption, is the endpoint of this process.

However, a modern Darwinian perspective would also predict that routine but low-level consumption of alcohol can be beneficial to humans. Typical physiological levels of exposure (as well as the extremes) are hard to estimate statistically for the average drinker, and there is also a high level of variability among individuals. But drink we certainly do, although perhaps many of us now consume only up to the levels of blood-alcohol concentrations reached during the ingestion of fermenting fruit. As expected from evolutionary interpretations of hormesis, and as shown empirically in epidemiological studies, some chronic exposure to alcohol is beneficial to health. More generally for humans, the consumption of fats, animal protein, and sugar can be viewed as natural dietary behaviors, but only within a limited range of exposure. Elevated and historically abnormal rates of ingestion, by contrast, result in a host of metabolic problems. The modern quest for better diets can thus be firmly anchored within the burgeoning field of evolutionary medicine, given this deep-time view of many of our favorite food items.

AN EVOLUTIONARY HANGOVER?

If contemporary responses to alcohol are conditioned by the dietary ghosts of frugivory past, then a number of predictions can be made about the feeding and foraging biology of fruit-eating animals, including many primates. It is remarkable how little is known about the natural occurrence of alcohol and its varied roles in dietary behaviors. In part, the comparative biology of alcohol exposure and response (a field we can term ethanology) has been neglected because overt inebriation is quite rare in the animal kingdom. If drunkenness were routine, then it would certainly have been well-studied by now. Instead, animals rarely if ever have the opportunity to feed on alcohol to excess, given that it can only be consumed at low concentrations within gut-filling fruit pulp. Exposure is thus slow and drawn out over the course of a feeding bout, and there is no chance for binge drinking. Given the potential costs of drunkenness for wild animals, it is also clear that selection would act to facilitate rapid clearance of the alcohol molecule, as well as mitigation of its inebriating effects. These predictions can all be tested. Using the tools of modern behavioral biology, comparative evolutionary methods, and genomic analyses to characterize DNA sequence variation underlying relevant physiological pathways, the historical signatures of selection on alcohol-related responses can be discerned.

First, however, we have to determine how animals might find alcohol-containing fruits in the wild. For fruit flies, the answer is clear. An upwind flight once they smell the alcohol molecule is usually sufficient to take them to fermenting fruits, although this behavior has only been analyzed under laboratory conditions. But can birds and mammals do the same thing? And what is the olfactory sensitivity of these animals to alcohol and the other odors of ripe and fermenting fruit (including such fragrant compounds as esters and acetic acid)? We know that some monkeys can taste alcohol and can presumably smell it as well, but do they use this cue to find fruit over long distances? Field experiments would be difficult in this regard, although the technologies of remote telemetry and miniaturized GPS tracking of free-ranging animals are becoming increasingly sophisticated. It would also be important to measure the time-varying alcohol signal in the environment, which is a complicated three-dimensional task. To this end, portable gas chromatographs could be used to obtain point samples of atmospheric alcohol vapor at various distances from a fruiting tree. Winds vary in both space and time, however, and it would be best to correlate displacement of moving animals with local alcohol concentrations in the air. Alternatively, artificial alcohol plumes could be generated in the field to try to elicit behavioral changes in monitored animals. We do inadvertently carry out this experiment when we open a beer bottle outside and bring in the fruit flies. It would be fascinating to conduct similar trials within tropical rainforests to try to attract both mammalian and avian fruit eaters.

One experimental means of testing attraction to alcohol-containing resources would be to make artificial fruits of agar or gelatin and to deploy them outdoors at feeding stations. Such fake fruits could be filled with alcohol solutions of different concentrations (ecological Jello shots, if you will) and could also be artificially colored using food dye. In addition to the possible use of smell over long distances, many animals use vision to find ripe fruit. Interaction between smell and color is a possibility, with the former cue working best over long distances and the latter only close up, particularly within visually obstructed habitats. And once nearby, animals may smell individual fruits to assess their ripeness, flavor, and alcohol content. All of these behaviors could be recorded with small video cameras at feeding stations. From the ensuing films, the reactions of individuals could be studied as they approached the station and then as they discriminated among potential food choices, ultimately choosing one for consumption.

Once animals select and start eating real fruit, what amounts of alcohol are they actually ingesting? We have surprisingly little information about the concentrations within ripe fruit, and also about the total amount consumed during a feeding bout. One obvious prediction would be that fruit eaters in the wild preferentially find and consume fruits with non-negligible levels of alcohol, as such fruits will reliably signal ripeness and the presence of sugars. It will be of particular interest to assess patterns of alcohol buildup during the often lengthy ripening sequence. Fermentation may happen early on in development, as yeast spores can land on flowers and then encapsulate within fruit as they grow and mature. Once simple sugars become available within the pulp, any yeasts present are free to ferment at will. Abrasion of the fruit's surface at various stages during growth also can permit yeast spores to germinate and subsequently multiply in the pulp's periphery. Some fruits naturally fall to the ground, others are knocked off the shrub or tree in question; ensuing tissue damage may facilitate subsequent microbial activity, including the growth of yeasts. The natural alcohol concentrations deriving from these different modes of yeast entry and growth within fruit are unknown. In the field, the alcohol content of homogenized fruit pulp can be measured directly using a portable infrared analyzer. This device characterizes specific kinds of organic compounds within a small liquid sample and can differentiate ethanol from other kinds of alcohols and organic compounds produced by yeast.

At some point, the fermentation of fruits is followed by substantial rot and decay. Innumerable bacteria are the guilty party, coming into play after yeasts have consumed sugars and produced alcohol. Badly decayed fruits appear to be avoided by animals in the wild, but at this stage there is probably little alcohol left amid the microbial rot. Humans tend to avoid rotting fruit as well, but we also select carefully to avoid eating unripe fruits. Somewhere in the middle of the ripeness spectrum is, literally, a sweet spot that we prefer. Low-concentration alcohol in aqueous solution is obviously desirable to us as well, but are fruits with small amounts similarly tasty? Some of these dietary preferences must be culturally acquired in humans. But we can also more ask generally, what role does alcohol play in the selection of ripe fruits by ourselves and other primates? And what levels of alcohol are typically detectable? The arguments presented in this book suggest that all fruit eaters should have fairly low taste thresholds for alcohol. Unfortunately, the relevant physiological experiments have only been carried out for a few species. Fruit eaters clearly use a variety of different cues to assess fruit palatability, and these will be challenging to disentangle for a complex task such as food selection. Nonetheless, carefully designed experiments to assess the effects of alcohol concentration on fruit choice are definitely possible, even under field conditions.

At this stage, we simply don't know what typical ripening looks like relative to either alcohol content or microbial populations, and we certainly don't have any information on how this may influence eating behavior by frugivorous animals. These questions represent conceptually low-hanging fruit and are ripe for the picking by enterprising field ecologists. For example, bacterial and fungal populations within fruit pulp can be quantified using standard microbiological methods. It is also possible to obtain accurate measurements of fruit color (using a device called spectroradiometer) as well as data on softness and other mechanical properties (which tend to be correlated with suitability for consumption). Particularly important here will be accurate and quantitative characterization of the widely used terms "ripe," "over-ripe," and "rotten." Such words are often used by ecologists to describe particular conditions of fruits, but we know very little about the kinds of botanical variation that these terms encompass, or about what these might mean for the feeding responses of different animals. Human perceptions of fruit edibility, for example, may deviate dramatically from those of wild primates.

We also have no information on the physiological concentrations of alcohol that are attained by animals in the wild. Behavioral changes associated with alcohol consumption are similarly unstudied, with the exception of the anecdotal accounts referred to in chapter 3. Other than taking blood samples and getting measurements of the associated alcohol concentrations, it is hard to imagine how one might systematically screen wild frugivores to determine typical values of exposure. Feeding stations, however, might provide one means of indirectly assessing physiological concentrations. As with humans operating Breathalyzers, it may be possible to sample the exhaled breath of an animal as it accesses fruit within a mask-type configuration. Theoretically, alcohol content in such a breath can be correlated with that in the blood, although this calibration is very sensitive to species identity, body mass, gender, and other confounding biological factors. Behavioral responses to alcohol ingestion, including possible effects on social interactions, will be similarly challenging to assess, although these could potentially be studied using artificial fruits presented at outdoor stations, along with remote video monitoring.

Naturally occurring alcohol may serve additional ecological functions that, to date, have received little attention. Animals consuming fruits may be indirectly facilitating the dispersal of yeasts if their spores can survive travel through the digestive system or if the fruit is relocated and then abandoned. Fermentation within fruit, in addition to its role in microbial competition (chapter 2), would then be under positive selection if such dispersal were to enhance subsequent yeast growth and reproductive spread. And alcohol, by virtue of the aperitif effect in enhancing feeding, may indirectly enhance overall rates of seed consumption and dispersal by fruit-eating birds and mammals. This outcome would potentially benefit the plants in question, suggesting a further complexity of interactions in the evolutionary triangle among fruit, yeast, and animals. Although fundamental to the biology of fermentation, these and other evolutionary pressures acting on yeasts to produce alcohol, as well as its specific competitive effects on other microbes within fruits, are essentially unstudied in natural contexts.

In laboratory settings, animal preferences for booze can be investigated using liquid alcohol solutions that mimic human drinking behavior, as is classically done by addiction researchers. Alternatively, it is possible to use alcohol mixed with more solid nutritional substances that simulate fermenting fruit. We might ask, for example, how similar are the responses of lab rats drinking liquid booze at varying concentrations to its presentation within gelatin fruits that also provide nutritional rewards? Do fruit-eating animals prefer, for example, the fairly low levels of alcohol that are characteristic of natural fermentation when it is commingled with useful calories? And does this simultaneously act as a feeding stimulant? We can also ask how food consumption and gut filling influence the total amount of ingested booze, all the way up to the point of satiation. Can the urge to consume alcohol, for example, be ultimately overridden by a carbohydrate-rich and stomach-filling meal? For animals, just how tight is this proposed linkage between food and alcohol consumption? And do similar neurophysiological pathways of reinforcement underlie addiction to both food and alcohol? If so, then an evolutionary description of how these pathways came into being may yield insight into the control mechanisms involved in extreme consumption.

We know little about the life history implications of low-level alcohol consumption for anything other than humans and fruit flies (as discussed in chapter 3). Alcohol has potentially wide-ranging effects on varied aspects of animal health, and the study of long-term exposure for habitual fruit eaters would be informative. These experiments would be particularly difficult and expensive to carry out on primates but would provide an important link to results from epidemiology demonstrating major health benefits. The molecular mechanisms by which alcohol provides such advantages also merit further investigation. For example, the reduction in cardiovascular risk seen in humans cannot pertain to fruit flies (which possess a very different and much simplified type of circulatory system). But low-level exposure of these flies to alcohol nonetheless enhances their longevity, as well as the fecundity of the females. What exactly are the mechanisms underlying such pronounced consequences? And do comparable effects pertain to fruit flies in the wild? Or to other animals that are chronically exposed to low-concentration alcohol? The antimicrobial effects of alcohol potentially underlie these health benefits in all animals, including ourselves, and their role in deterring infections deserves further experimental attention.

Because some aspects of the evolutionary process can be reconstructed from DNA sequences, genomic studies examining the enzymes involved in alcohol metabolism may illustrate patterns of historical exposure. Substantial genetic variation in different species of fruit flies and among different contemporary human populations (chapter 6) should be mirrored in other kinds of animals exposed to different dietary levels of alcohol. For example, the predominantly fruit-eating lowland gorillas would be predicted to have fast-acting ADH and ALDH enzymes in comparison with their mountain-dwelling counterparts (which only rarely eat fruit; see figure 4). Specialized frugivorous birds tend to have faster ADH enzymes and to metabolize alcohol more quickly than more omnivorous species. We would expect to see similar patterns of variation when we compare fruit-eating bats (e.g., the Old World fruit bat family Pteropodidae and many representatives of the New World phyllostomids; see plate 11) with closely related bats that specialize on insects and other nutritional resources containing no alcohol.

Molecular studies of other groups of fruit-eating mammals and birds may also be informative. In Central and South America, for example, two species of carnivore in the raccoon family (namely, kinkajous and olingos) live only in trees and feed primarily on fruit. We would accordingly predict that their enzymes involved in alcohol degradation are very different from those of other carnivores, which are mostly meat eaters (e.g., the cats) or omnivores (e.g., North American raccoons). Carnivorous mammals are particularly interesting subjects for study in this regard. Over evolutionary time, many of them have functionally lost their sweet taste receptors given their meat-based diet, and they behaviorally avoid sweet-tasting compounds. Fruit-eating carnivores, by contrast, would be predicted to have retained the associated taste genes and to prefer sweet flavors by virtue of their association with ripe fruit. Polar bears, for example, eat only meat, fat, and carrion, whereas grizzly bears seasonally include many small ripe fruits in their diet. A broad phylogenetic comparison of all the carnivores would be informative at the levels of sugar preference and alcohol metabolism.

More generally, the species-rich tropical rainforests provide a number of advantages for evaluating the natural role of alcohol in animal foraging ecology. The number of fruiting plant species in these forests worldwide is impressively high, on the order of tens of thousands. Each of these exhibits a unique set of fruit colors, flavors, ripening behavior, seasonal abundance, and other ecological aspects relevant to potential consumers. Seasonal variability in ripe fruit can also be high. With greater distances either north or south from the equator, a dry season up to four months long can result in fruit scarcity and associated behavioral responses by animals. The taxonomic and physiological diversity of fermenting yeasts in the tropics is badly understudied, but it must influence alcohol production within fruit and its corresponding signal to frugivores. Overall, such biological richness in both plants and yeasts suggests the potential for diverse outcomes in animal mechanisms of fruit identification, preference, and consumption relative to alcohol content. Also within tropical and subtropical environments, high species richness in birds, insects, and mammals suggests comparable variation in behavioral and metabolic responses to alcohol. Some of these responses may predispose particular species to excessive consumption when given the opportunity.

We might predict, for example, that higher levels of natural exposure will yield more pronounced addictive responses to any abnormal availability of alcohol, either in liquid form or when mixed with more solid food. Similarly, a genetic capacity to quickly metabolize the alcohol molecule could be correlated with the tendency to consume to excess, as occurs in humans. However addictive responses might be defined, including such features as increased tolerance and the presence of withdrawal symptoms, they ultimately must derive from molecular pathways and sensory biases intrinsic to the brain. And brain evolution, with its complex and varied sensory capacities, has been molded by selection pressures in the past. Broad comparative studies of feeding by different frugivorous species may reveal similarities in those mechanisms used to sense and respond to the alcohol molecule. Its obligate association with dietary calories provides the physiological framework for psychoactive and energetically beneficial rewards. Unfortunately, this association can also pave the way to excessive consumption. If behaviors similar to what we term to be addiction in humans are found in a number of different fruit-eating species, then a strong evolutionary basis for addiction to alcohol could reasonably be inferred. More broadly, many other addictive substances used by humans may act similarly on these reward pathways. Recreational drugs, in other words, may co-opt what once was a beneficial outcome to yield self-reinforcing and maladaptive behaviors. Although tragic today, our chemical addictions may simply reflect a more gentle evolutionary past of primates enjoying the tastes and sensations of ripe fruit in tropical rainforests.

IN VINO VERITAS

This Latin saying, often attributed to Pliny the Elder (who actually translated the original from the Greek poet Alcaeus), refers to the ability of wine to loosen tongues and to reveal the truth. More broadly, we obviously have a special relationship with alcohol. Innumerable cultural and social activities, from conversation to artistic creation and copulation, are accompanied by and facilitated by drinking. Often consumption proceeds to excessive levels. The *Dictionary of American Slang*, for example, contains far more words for "drunk" than for any other adjective. Many of these are humorous and indirectly acknowledge the widespread role of alcohol in facilitating social interactions. By way of extreme example, the contemporary magazine *Modern Drunkard* (no, I am not making this up; see *drunkard.com*) regularly publishes both informational and simply entertaining stories about the social culture of alcohol. Such a comical perspective on drinking is perhaps best offset by viewing the website of the U.S. National Highway Traffic Safety Administration. Here, one can learn directly about fatality rates associated with alcohol-impaired driving, which results, like clockwork, in one death every fifty minutes throughout the year. Between these extremes reside most individuals who use alcohol responsibly and who accordingly obtain some measure of health benefit from it. Can there be a middle ground corresponding to the safe consumption of alcohol?

One approach to regulating alcohol, of course, is simply to try to prohibit it altogether. In the United States, the history of such efforts provides fascinating insights into the intrinsic difficulties of such public policy. Prohibition efforts began in fits and starts in the nineteenth century and were formally initiated nationwide in 1919. From the beginning, law enforcement was simply incapable of preventing small-scale production and distillation operations, although larger industrial concerns were essentially shut down. Such disruption was, predictably, offset in part via widespread bootlegging across state lines and from Canada. Newspaper reports of routine alcohol consumption by numerous senators and congressmen did little to instill confidence in federal prohibition efforts. Medical exemptions were also permitted; none other than Winston Churchill, after being hit by a car in New York City in 1932, was formally prescribed a "naturally indefinite" quantity of alcoholic spirits, "especially at meal times." Prohibition was repealed in 1933, ending more than a decade of reduced consumption but hardly the elimination of alcohol. Today, at the national level, only a number of Muslim countries impose strict legal prohibitions on the production and consumption of alcohol-containing beverages, although reportedly these efforts are often circumvented. And it is certainly possible to publicly enjoy alcohol in various majority-Muslim countries located on the periphery of the core Islamic world (e.g., Malaysia and Morocco). If humans possess an intrinsic physiological drive to consume alcohol, albeit in limited quantities, then the enforcers of major cultural restrictions against booze have their work cut out for them.

For example, in the Muslim quarter of the Chinese city of Xi'an I have often enjoyed spicy lamb skewers, along with grilled flatbread and what would seem to be a natural accompaniment, namely beer. But curiously, the restaurants and their proprietors refuse for religious reasons to serve alcohol. Instead, they simply place the drinks order with a nearby restaurant run by ethnic Han Chinese and then serve up beer in aluminum or ceramic teapots, as if to conceal the identity of the drink from any observing deity. Maybe this works for the higher power, and in the meantime restaurant commerce continues unimpeded. On a broader national scale in the United States, the ongoing War on Drugs (kicked off by the drinker Richard Nixon in 1971) conveniently omits alcohol from its purview. It's not that alcohol isn't a potentially addictive and dangerous inebriant, but rather that it's just too close to home. Too many of us simply seem to enjoy it safely, independent of any broader societal costs that are attributable to others. Why mess around with an apparently natural drug that makes so many of us happy, facilitates so much of our cuisine and our social lives, and contributes to the national economy?

Patterns of federal funding for alcoholism research in the United States indicate similarly conflicted views. For example, there is no particular consensus as to what constitutes an addictive drug and how the government might systematically study addiction. The National

Institutes of Health currently comprise twenty-seven institutes, of which one is the National Institute on Drug Abuse, which covers addictive disorders associated with pharmacological agents. But, puzzlingly, another institute is called the National Institute on Alcohol Abuse and Alcoholism, a designation obviously posing substantial overlap with the broader study of drug abuse. Is not alcohol also an addictive substance? In 2012, administrative plans were afoot to merge these two institutes into a single entity, following their approximately forty years of independent existence. But the mere fact that alcohol was decoupled administratively from all other abused drugs speaks volumes as to the dually positive and negative features of drinking. Some of our problems in formally identifying the disease of alcoholism necessarily derive from the fact that moderate drinking can be healthful. The tipping point towards abuse remains unclear, in part because we have failed to recognize the beneficial aspects of natural alcohol exposure in many animals, both human and otherwise.

As a consequence, policy approaches towards controlling alcohol consumption typically have limited success. It is very difficult to identify and limit dangerous behaviors while at the same time permitting moderate and safe levels of drinking. In this respect, a comparison with gun control measures (or more accurately, their absence) in the United States is informative. Tens of millions of gun owners nationwide responsibly own and operate firearms, but a relatively small fraction abuse the privilege and engage in criminal behavior. Numerically, however, the absolute number of such abusers runs into the hundreds of thousands. As currently practiced, gun ownership in the United States statistically and inevitably results in the deaths of about thirty thousand American citizens annually, including many children. Can we eliminate this latter outcome but still permit the regulated use of firearms? Technology and governmental intervention would certainly help in this regard (e.g., registration of guns, bullet microstamping, high taxes on gun and ammo production, and so on). Similarly, some level of technological intervention might limit extreme consumption of alcohol (e.g., the use of implantable drug devices for the delivery of disulfiram, and clever means of breath analysis to prevent drunks from driving). But thus far, such measures have not been promising in the treatment of alcoholism or in the mitigation of its tragic consequences.

If the desire for alcohol is indeed hard-wired in humans, then legal restrictions as well as informational campaigns to increase awareness of associated hazards simply may not work. Certainly on college campuses in the United States, alcohol remains highly attractive in spite of extensive efforts to inform students of the dangers of excessive drinking. Raising legal age limits for drinking is also easily skirted, both in private places and at bars where the phenomenon of the false ID is widespread. In the United Kingdom, a burgeoning culture of binge drinking over the last two decades has resulted in dramatically increased rates of public drunkenness, associated violent behavior, and a series of government efforts to regulate consumption. The tone of many of the associated reports borders on desperation. And worldwide, we continue to tolerate broad commercial dissemination of an uninhibited drinking culture, complete with widespread advertising and the association of positive role models with alcoholic beverages. The power of the marketplace, with profits pouring in for shareholders of companies, can in the short term effectively dominate those public health concerns that materialize over the course of decades. In the United States and elsewhere in industrialized countries, commerce has clearly trumped the public interest when it comes to the question of alcohol promotion among the drinking populace.

Perhaps the best policies for controlling alcohol consumption are the physical restriction of its availability, along with taxes that increase prices for the end user (but not to levels so high as to create a large illicit market in home brew). Such approaches, along with informational campaigns, have worked well over the last forty years in reducing smoking incidence in the United States. They impose direct costs on consumers and create a financial deterrent for those whose behavior damages the health and well-being of others. Externally imposed limits to availability, in other words, are more likely to be effective in the long run than attempts to elicit self-policing of drinking behavior. Because booze is relatively cheap (with rock-bottom beer and wine being less costly than an equivalent volume of bottled water), ample scope exists to increase the tax burden on alcohol, and thus the costs of its consumption. Although at some point such a strategy may promote a switch to home-brewing and non-payment of taxes, we are currently nowhere near such a threshold. Excise taxes on alcoholic beverages have actually experienced a relative decline with respect to retail prices over the last several decades in the United States. This is one public policy approach that clearly deserves further consideration, particularly given that the associated revenue can be used to help mitigate the adverse effects of alcohol consumption. Enhanced technological methods to deter drunk driving (such as car ignition devices that require an alcohol-free breath to start the vehicle, followed by occasional further breaths by the driver once the car is in motion) would be also desirable. Similarly, further lowering of legal blood-alcohol limits would reduce the incidence of this particularly dangerous behavior. Although per capita fatality rates associated with drunk driving have generally been in decline across the United States in the last twenty years, there is still substantial room for further improvement.

More generally, identification of alcoholism as a disease of nutritional excess suggests potentially common strategies for managing a number of addictive behaviors. As with alcohol abuse, ongoing epidemics in diabetes and obesity are consistent with high and historically anomalous rates of consumption. The immediately obvious comparison for alcoholism is the contemporary problem of excessive sugar intake, particularly in the form of high-fructose corn syrup, which is added to many processed foods. Note also that the word "fructose" is derived etymologically from the Latin *fructus,* meaning fruit. This form of sugar is mostly found in wild rather than in domesticated fruits, and in comparison to glucose actually tends to stimulate the appetite. The biochemical differences between all sugars and alcohol (i.e., the ethanol molecule) are small, as the latter derives from the fermentation of these simple carbohydrates. It is therefore not surprising that many physiological problems associated with high-level exposure to both compounds are similar. In aggregate, these have been termed the metabolic syndrome, and they include such unfortunate medical consequences as insulin resistance, cardiac problems, pancreatitis, and liver dysfunction. The economic impacts of the metabolic syndrome are also comparable in magnitude to those associated with alcoholism. Regulation of sugar and alcohol availability to human populations, via controls on both production and purchase, may be the only effective means of interfering with our otherwise insatiable appetite for calories. Such eating urges are eminently sensible in evolutionary terms but can quickly lead to problems when cheap food and booze are produced industrially and widely distributed. Curbing availability of addictive substances may ultimately be much more effective than trying to interfere with intrinsic physiological drives as shaped by millions of years of human evolution.

A number of years ago, I ran a freshman seminar course at Berkeley which was devoted to the drunken monkey hypothesis. When I asked the students why they liked to drink alcohol (and only some of them did, whereas all of them must have been below the legal drinking age), invariably the answer was some version of "because it tastes good." This response must of course be true at some basic level. It doesn't address, however, the more complicated question of why our taste receptors and assessment of flavor have evolved to render certain chemical compounds preferable and others disagreeable. Why, for example, do we not regularly drink vinegars of different vintages and compositions? Why the obsession with the immediate alcoholic product of fermentation? And why its excessive use? In evolutionary biology, an important distinction is drawn between proximate and ultimate causes. The former refer to the immediate physiological, environmental, and behavioral factors influencing an outcome, whereas the latter indicate those longer-term selective pressures responsible for determining the relative strength of these factors over many generations. Here I have emphasized that the short-term responses underpinning alcohol preference and addiction reflect biases of a brain that has been molded and formed over evolutionary timescales. If we are to search for a cure to the disease of alcoholism, then we must also recognize that today's nutritional environments are very different from those of ten thousand years ago, let alone ten million years ago.

When we think about alcoholism, it is important to realize that convincing explanations for such a powerful attraction to booze have been wanting. Weakness of character is a classic interpretation that, in reality, tells us nothing informative about either the disease or its sufferers. The inability to control drinking must certainly reflect some feature of impulse control, but the alcohol molecule itself must be involved as well. If alcohol, in its diverse and potent forms, activates pre-existing and once useful motivational biases, then in essence we are today abused by the molecule, rather than abusing it ourselves. The medical profession long ago abandoned the abuse concept of alcoholism, given its negative connotations and blame imputed to the patient. Instead, modern practice considers this disease to be just like hundreds of other behavioral disorders, a medical condition for which sufferers require treatment. If reward pathways in our brain are indeed exploited by alcohol to yield false perceptions of caloric reward, then the behavioral pressures motivating consumption cut to the core of the daily physiological survival of the organism. Psychological, social, or philosophical explanations of alcoholism, however well intentioned, will be of limited utility here.

For centuries, the prevailing view of drug addictions has been that such behaviors, along with language and consciousness, are unique to humans. As a consequence, both analysis and treatment of responses to addictive drugs (including alcohol) have been conceptually decoupled from the natural environments within which our behavior and physiology evolved. A key feature of many addictions, namely their potential evolutionary advantage, is thus ignored. I have argued throughout this book for a deeper-time perspective on alcohol, and for one that acknowledges both the positive and negative consequences of routine drinking. As a medical problem, alcoholism has been persistent across human societies and has defied systematic characterization, let alone effective treatment. Ultimately, only an evolutionary perspective can fully decipher our complex and ambiguous responses to the alcohol molecule.

Postscript

> When I read about the evils of drinking, I gave up reading.
> Henny Youngman

Drink in hand on Barro Colorado Island (and in emulation of the bonobo portrayed in plate 12), I contemplate our conflicted relationship with alcohol. On the one hand, drinking can positively enhance many aspects of our social and personal lives. Alcohol can also result in substantial health benefits when consumed in moderation. On the other hand, excessive drinking ruins lives over the long run, including those of the alcoholic and her or his family members, along with the innumerable victims of drunk driving and other senseless acts committed while under the influence. In his comical yet insightful statement, Henny Youngman clearly embraces the pleasures of drinking but also indirectly acknowledges their negative consequences. This is the double-edged sword of alcohol, and one that makes the public health job of prescribing reasonable and safe drinking behavior very difficult.

Am I drinking too much? Or possibly too little? I sometimes ask myself these questions, both as the son of an alcoholic and as someone who well appreciates beer, wine, and distilled spirits (albeit in careful moderation). When I first started developing the drunken monkey hypothesis in the late 1990s and began to read about the epidemiology of alcohol exposure, it quickly became clear that my own consumption was statistically on the low side relative to potential health benefits. But would drinking a little more every day also increase the chances that I would transition into more dangerous patterns of drinking? Would I start to drink and drive, for example? And was there any way to predict these possibilities in a scientific way? Based on the published literature and consensus of the medical profession to date, it is clear that there can be no simple or definitive answer to these questions. Given how little we know about the disease of alcoholism, it is now obvious to me that individuals just have to gauge this situation for themselves, using different kinds of evidence (including gender, family background, age, and other factors) in consultation with their physicians. And, simply put, we also just have to hope for a good outcome relative to our long-term drinking behaviors. Statistically speaking, a good outcome will in fact be the case for most people who drink regularly. But this assessment can be of no consolation to those who do end up as alcoholics, and to others who tragically suffer indirectly from the consequences.

To broaden the scientific scope of the issues now posed in this book, I and my friend and colleague Michael Dickinson organized in 2004 a symposium to address the varied and fascinating biological aspects of alcohol exposure. In classical Greek, the word "symposium" indicates a drinking party held for the purposes of scholarly discussion. Our event, as part of the annual meeting of the Society for Integrative and Comparative Biology, was fortuitously held in New Orleans that year, where we had ample drinking opportunities to celebrate the etymological origins of the word. A full day of talks was followed by dinner at one of New Orleans' celebrated restaurants, with suitable libation accompanying the fine regional cuisine. More generally, the talks in the symposium covered a diversity of topics on the comparative biology of alcohol exposure, ranging from fruit flies to the health consequences of drinking in humans. Above all else, the ensuing physiological and evolutionary questions that were generated by the symposium convinced me that far too much focus has been placed (albeit understandably) by medical researchers on addiction-related physiology and the behavior of model animal systems such as those involving mice and rats. And the use of rodent analogs for drinking behavior in modern humans is, moreover, unlikely to tell us much about the natural biology of our attraction to alcohol. The fundamental lack of progress in treating the disease of alcoholism reflects these biases and ultimately underscores the failure to develop a broader and comparative biology of alcohol exposure.

Over the past decade, such neglect has been partially addressed by expanding studies that use fruit flies as a model system for understanding the molecular underpinnings to both inebriation and addiction. The natural foraging ecology of different vertebrate species relative to alcohol within fruit and nectar has also received increased attention, as exemplified by the treeshrews discussed in chapter 3. But of the animal species potentially exposed to alcohol in rainforests and other environments where fermenting yeasts can be important, only a tiny fraction have been identified, let alone studied in detail as to how and why they respond to alcohol. Deeply embedded within our genome and those of other species are important clues as to why we drink today, both in moderation and to excess. We owe it to the sufferers of alcoholism, and to those who indirectly endure the outcomes of this disease, to pursue these questions further.

SOURCES AND RECOMMENDED READING

1. INTRODUCTION

The literature on alcohol and alcoholism is immense. The NIH MedlinePlus website on alcohol (www.nlm.nih.gov/medlineplus/alcohol.html) provides the easiest access to the primary medical literature. A good cultural and biomedical overview of alcohol use and abuse is that of Griffith Edwards (*Alcohol: The World's Favorite Drug,* 2003, St. Martin's Griffin, New York). A popular-level introduction to the field of evolutionary medicine is provided by Ralph Nesse and George Williams in their book entitled *Why We Get Sick: The New Science of Darwinian Medicine* (1996, Vintage Books, New York). Genotypic diversity in modern humans, including many of the associated consequences for health, is covered in detail in the multi-author volume entitled *Human Evolutionary Biology* (2010, Cambridge University Press, Cambridge). The drunken monkey hypothesis for human alcoholism was first published in the *Quarterly Review of Biology* ("Evolutionary origins of human alcoholism in primate frugivory," 2000, 75:3–15). The astonishing fruit-eating fishes of the Amazon river basin are described in Michael Goulding's book *The Fishes and the Forest* (1980, University of California Press, Berkeley).

2. THE FRUITS OF FERMENTATION

The classic treatment of the diversity of fruiting plants is that of Henry Ridley (*The Dispersal of Plants Throughout the World,* 1930, L. Reeve, Ashford, Kent). A modern and well-illustrated treatment of fruit biology can be found in the book by Wolfgang Stuppy and Rob Kesseler entitled *Fruit: Edible, Inedible, Incredible* (2008, Firefly Books, Buffalo, NY). A basic introduction to frugivory in the broader context of plant-vertebrate interactions is provided by Carlos Herrera and Olle Pellmyr in their book *Plant-Animal Interactions: An Evolutionary Approach* (2002, Blackwell Science, Malden, MA). The symposium proceedings entitled *Seed Dispersal and Frugivory* (2002, CABI Publishing, Wallingford, UK) also provide a good introduction to the primary literature and ongoing research questions. The biology of yeasts and their natural ecology, including association with fruits and interactions with invertebrate dispersers, are well covered in the multi-author handbook entitled *Biodiversity and Ecophysiology of Yeasts* (2006, Springer, Berlin). The biology of alcohol production by yeasts is treated by Christopher Boulton and David Quain in *Brewing Yeast and Fermentation* (2001, Blackwell Science, Oxford). Interactions between fruits and microbes are comprehensively reviewed by Martin Cipollini and Edmund Styles in the series *Advances in Ecological Research* ("Relative risks of microbial rot for fleshy fruits: significance with respect to dispersal and selection for secondary defense," 1992, 23:35–91). A detailed description of the relative tolerances of yeast and bacteria to alcohol is that of Lonnie Ingram and Thomas Buttke in the series *Advances in Microbial Physiology* ("Effects of alcohols on micro-organisms," 1984, 25:253–300). Experimental measurements of bumblebee foraging responses to yeasts within floral nectar were recently published by Carlos Herrera and colleagues in *Ecology* ("Yeasts in nectar of an early-blooming herb: sought by bumble bees, detrimental to plant fecundity," 2013, 94:273–279). I discuss ripening profiles and alcohol content of palm fruits in the paper entitled "Ethanol, fruit ripening, and the historical origins of human alcoholism in primate frugivory" (*Integrative and Comparative Biology*, 2004, 44:315–323).

3. ON THE INEBRIATION OF ELEPHANTS

The potential for elephant drunkenness is discussed by Steve Morris and co-authors in the journal *Physiological and Biochemical Zoology* ("Myth, marula, and elephant: an assessment of voluntary ethanol intoxication of the African elephant (*Loxodonta africana*) following feeding on the fruit of the marula tree (*Sclerocarya birrea*)," 2006, 79:363–369). Inebriated birds are described in a paper in *Avian Diseases* ("Suspected ethanol toxicosis in two wild cedar waxwings," 1990, 34:488–490), and in a more recent paper in the *Journal of Ornithology*

("Strong circumstantial evidence for ethanol toxicosis in Cedar Waxwings (*Bombycilla cedrorum*))," 2012, 153:995–998). William Miller reviews cases of drunken Lepidoptera, including butterflies, in the paper entitled "Intoxicated lepidopterans: how is their fitness affected, and why do they tipple?" (*Journal of the Lepidopterists' Society*, 1997, 51:277–287). Responses of fruit bats to alcohol, along with its natural levels within fruit, are described by Francisco Sánchez and colleagues in the journal *Behavioural Processes* ("Ethanol ingestion affects flight performance and echolocation in Egyptian fruitbats," 2010, 84:555–558). The night-blooming Malaysian palm with alcohol-bearing nectar, along with physiological assessment of its consumption by animal pollinators, was described by Frank Wiens and others in 2008 ("Chronic intake of fermented floral nectar by wild treeshrews," *Proceedings of the National Academy of Sciences USA*, 105:10426–10431).

A classic introduction to fruit fly biology is that of Milislav Demerec (*Biology of Drosophila*, 1950, Wiley, New York). A more popular account is given by Martin Brooks in *Fly: The Unsung Hero of Twentieth-Century Science* (2001, HarperCollins, New York). The use of fruit flies in molecular studies of susceptibility to alcohol is reviewed by Ulrike Heberlein and colleagues in *Human Genetics* ("*Drosophila melanogaster* as a model to study drug addiction," 2012, 131:959–975). Olfactory responses of fruit flies to natural fermentation products are described by Ary Hoffman and Peter Parsons in the *Biological Journal of the Linnean Society* ("Olfactory response and resource utilization in *Drosophila*: interspecific comparisons," 1984, 22:43–53). Alcohol and sex deprivation in fruit flies are the topics of a fascinating paper in *Science* by Galit Shohat-Ophir and colleagues ("Sexual deprivation increases ethanol intake in *Drosophila*," 2012, 335:1351–1355). The fermentation odors produced by Solomon's lily are described by Johannes Stökl and colleagues in *Current Biology* ("A deceptive pollination system targeting drosophilids through olfactory mimicry of yeast," 2010, 20:1846–1852).

A good introduction to hormesis is the book by Mark Mattson and Edward Calabrese entitled *Hormesis: A Revolution in Biology, Toxicology and Medicine* (2010, Springer, New York). General implications of hormesis for evolutionary biology were reviewed by Peter Parsons in the *Quarterly Review of Biology* ("The hormetic zone: an ecological and evolutionary perspective based upon habitat characteristics and fitness selection," 2001, 76:459–467). The role of alcohol in defense against parasites was recently described by Todd Schlenke and colleagues in *Current Biology* ("Alcohol consumption as self-medication against blood-borne parasites in the fruit fly," 2012, 22:488–493) and in *Science* ("Fruit flies medicate offspring after seeing parasites," 2013, 339:947–950). Therapeutic effects of natural alcohol exposure for a nematode were recently described by Paola Castro and colleagues in *PLoS ONE* ("*Caenorhabditis elegans* battling starvation stress: low levels of ethanol prolong lifespan in L1 larvae," 2012, 7:e29984). A basic introduction to the beneficial consequences of low-level alcohol consumption in humans was published by Art Klatsky in *Scientific American* ("Drink to your health?," 2003, 288:74–81). Two recent large-scale analyses of the effects of alcohol on heart disease and mortality are those by Paul Ronksley and colleagues ("Association of alcohol consumption with selected cardiovascular disease outcomes: a systematic review and meta-analysis," *British Medical Journal*, 2011, 342:d761) and by Micael Roerecke and Jürgen Rehm ("The cardioprotective association of average alcohol consumption and ischaemic heart disease: a systematic review and meta-analysis," *Addiction*, 2012, 107:1246–1260).

4. APING ABOUT IN THE FOREST

Good general introductions to the tropical rainforest are those by Richard Corlett and Richard Primack (*Tropical Rain Forests: An Ecological and Biogeographical Comparison*, 2nd ed., 2011, Wiley-Blackwell, Oxford) and by Jaboury Ghazoul and Douglas Sheil (*Tropical Rain Forest Ecology, Diversity, and Conservation*, 2010, Oxford University Press, Oxford). Ted Fleming and colleagues discuss frugivore ecology in the tropics in the *Annual Review of Ecology and Systematics* ("Patterns of tropical vertebrate frugivore diversity," 1987, 18:91–109). The interesting evolutionary history of frugivory is reviewed by Ted Fleming and John Kress in the journal *Acta Oecologia* ("A brief history of fruits and frugivores," 2011, 37:521–530).

A general overview of primate biology and evolution can be found in John Fleagle's book entitled *Primate Adaptation & Evolution*, 3rd ed. (2013, Academic Press, San Diego). Diets and foraging strategies of primates are discussed in detail by Gottfried Hohmann in a book chapter entitled "The diets of non-human primates: frugivory, food processing, and food sharing" (pp. 1–14 in *The Evolution of Hominin Diets: Integrating Approaches to the Study of Paleolithic Subsistence*, 2009, Springer Science, Berlin), and by Joanna Lambert in "Primate nutritional ecology: feeding biology and diet at ecological and evolutionary scales" (pp. 512–521 in *Primates in Perspective*, 2nd ed., 2010, Oxford University Press, Oxford). Primate sensory biology in relation to foraging behavior, along with data on fruit-alcohol concentrations, are covered by Nate Dominy in the journal *Integrative and Comparative Biology* ("Fruits, fingers, and fermentation: the sensory cues available to foraging primates," 2004, 44:295–303). Neurophysiological responses to various alcohols are documented by Matthias Laska and Alexandra Seibt in the *Journal of Experimental Biology* ("Olfactory sensitivity for aliphatic alcohols in squirrel monkeys and pigtail macaques," 2002, 205:1633–1643).

Detailed treatments of human paleodiets, along with discussion of the numerous difficulties intrinsic to their reconstruction, can be found in the multi-author volumes edited by Peter Ungar (*Evolution of the Human Diet*, 2007, Oxford University Press, Oxford) and by Jean-Jacques Hublin and Michael Richards (*The Evolution of Hominin Diets: Integrating Approaches to the Study of Paleolithic Subsistence*, 2009, Springer Science, Berlin). A recent textbook of evolutionary medicine is that by Peter Gluckman and colleagues (*Principles of Evolutionary Medicine*, 2009, Oxford University Press, Oxford). A recent review is provided by Steven Stearns in the *Proceedings of the Royal Society of London* Series B ("Evolutionary medicine: its scope, interest and potential," 2012, 279:4305–4321). Two new and useful journals in the field are *Evolution, Medicine & Public Health* and the *Journal of Evolutionary Medicine*. Nutritional and medical problems deriving from the overabundance of food in industrialized countries are discussed by Marion Nestle and Malden Nesheim in their book *Why Calories Count: From Science to Politics* (2012, University of California Press, Berkeley).

5. A FIRST-RATE MOLECULE

The early archaeological records of beer and wine production are well covered in two recent books by Patrick McGovern (*Ancient Wine: The Search for the Origins of Viniculture*, 2007, Princeton University Press, Princeton; and *Uncorking the Past: The Quest for Wine, Beer, and Other Alcoholic Beverages*, 2009, University of California Press, Berkeley). Paleolithic brewing in the Near East and its potential links with cereal domestication and feasting are comprehensively reviewed by Brian Hayden and colleagues in the *Journal of Archaeological Methods and Theory* ("What was brewing in the Natufian? An archaeological assessment of brewing technology in the Epipaleolithic," 2013, 20:102–150). The culture, biology, and practical implementation of food fermentation are discussed in the wonderful book by Sandor Katz entitled *The Art of Fermentation: An In-Depth Exploration of Essential Concepts and Processes from around the World* (2012, Chelsea Green Publishing, White River Junction, VT). Asian origins of the technology of distillation, the invention of "frozen-out wine," and early production of high-concentration alcohol are detailed by Hsing-tshung Huang in volume 6 (Biology and biological technology), part V (Fermentations and food science) of the magisterial series entitled *Science and Civilisation in China* (2000, Cambridge University Press, Cambridge). Diverse cultural practices of drinking are abundantly described in the volumes edited by Mac Marshall (*Beliefs, Behaviors, & Alcoholic Beverages: A Cross-Cultural Survey*, 1979, University of Michigan Press, Ann Arbor), by Dwight Heath (*Drinking Occasions: Comparative Perspectives on Alcohol and Culture*, 2000, Routledge, New York), and by Thomas Wilson (*Drinking Cultures: Alcohol and Identity*, 2005, Berg Publishers, Oxford). Iain Gately provides an amusing history of the world as seen through the prism of alcohol in *Drink: A Cultural History of Alcohol* (2008, Gotham Books, New York). Craig MacAndrew and Robert Edgerton emphasize the important social context of drinking and behavioral responses to alcohol in their classic book entitled *Drunken Comportment: A Social Explanation* (1969, Aldine, Chicago). The World Health Organization publishes annually an online global status report for worldwide patterns of alcohol consumption, associated medical issues, and policy responses (www.who.int/substance_abuse/publications/global_alcohol_report/en/index.html). A good overview of the aperitif effect in humans is that by Martin Yeomans ("Effects of alcohol on food and energy intake in human subjects: evidence for passive and active over-consumption of energy," *British Journal of Nutrition*, 2004, 92:S31-S34).

6. ALCOHOLICS AREN'T ANONYMOUS

Addiction biology in general is well covered by Carlton Erickson (*The Science of Addiction: From Neurobiology to Treatment*, 2007, W.W. Norton, New York). General reviews of the varied effects of alcohol can be found in the edited volume entitled *Alcohol and Human Health* (2008, Oxford University Press, Oxford). An integrative review of the physiological mechanisms underpinning alcoholism is provided by Rainer Spanagel ("Alcoholism: a systems approach from molecular physiology to addictive behavior," *Physiological Reviews*, 2009, 89: 649–705). A detailed multi-author treatment of the disease can be found in the weighty three-volume set entitled *Comprehensive Handbook of Alcohol-Related Pathology*, edited by V.R. Preedy and R.R. Watson (2005, Academic Press, London). Hereditary components to alcoholism are discussed by Joel Gelernter and Henry Kranzler in a review in *Human Genetics* ("Genetics of alcohol dependence," 2009, 126:91–99). Population-level differences in the ability to metabolize alcohol are reviewed by Howard Edenburg ("The genetics of alcohol metabolism: role of alcohol dehydrogenase and aldehyde dehydrogenase variants," 2007, *Alcohol Research & Health*, 30:5–13). The remarkable diversity of treatments used historically for alcoholism is covered by William White in his 1998 book entitled *Slaying the Dragon: The History of Addiction Treatment and Recovery in America* (Chestnut Health Systems, Bloomington, IN). Correlations between a sweet tooth and alcoholism are reviewed by Alexey Kampov-Polevoy and colleagues in the journal *Alcohol and Alcoholism* ("Association between preference for sweets and excessive alcohol intake: a review of animal and human studies," 1999, 34:386–395). The Alcohol and Public Health website of the Centers for Disease Control and Prevention (www.cdc.gov/alcohol/) contains excellent information on the costs of drinking at both personal and societal levels. Barron Lerner's book entitled *One More for the Road: Drunk Driving Since 1900* evaluates the historical and tragic intersection between alcohol use and motor vehicles (2011, Johns Hopkins University Press, Baltimore). The history of ADH polymorphism in East Asia and its relationship to historical patterns of rice cultivation are discussed by Yi Peng and colleagues in the journal *BMC Evolutionary Biology* ("The ADH1B Arg47His polymorphism in East Asian populations and expansion of rice domestication in history," 2010, 10:15). Modern experimental approaches using nonhuman primates to elucidate human drinking behavior are reviewed by Kathleen Grant and Allyson Bennett in *Pharmacology & Therapeutics* ("Advances in nonhuman primate alcohol abuse and alcoholism research," 2003, 100:235–255). The relevance of rodent models for understanding alcohol dependence in humans is critically assessed by John Crabbe in the journal *Genes, Brain, and Behavior* ("Translational behaviour-genetic studies of alcohol: are we there yet?" 2012, 11:375–386).

7. WINOS IN THE MIST

The Darwin Correspondence Project (www.darwinproject.ac.uk) and its associated physical publications permit direct access to thousands of fascinating letters both to and from Charles Darwin. Staffan Lindeberg comprehensively reviews linkages between ancestral nutritional strategies and human disease in his book *Food and Western Disease: Health and Nutrition from an Evolutionary Perspective* (2010, Wiley-Blackwell, Oxford). The evolutionary loss of sweet taste in carnivores is assessed by Peihua Jang and colleagues in a fascinating paper that links feeding preferences to the genetic modification of taste receptors ("Major taste loss in carnivorous mammals," *Proceedings of the National Academy of Sciences USA*, 2012, 109: 4956–4961). The intriguing history of prohibition efforts in the United States is detailed by Daniel Okrent in *Last Call: The Rise and Fall of Prohibition* (2010, Scribner, New York). Cogent arguments for the regulation of access to sugar and for its designation as an addictive substance have been made recently by Robert Lustig and colleagues in *Nature* ("The toxic truth about sugar," 2012, 482:27–29). Obesity and the metabolic syndrome relative to modern diets are also covered by Robert Lustig in his popular-level book entitled *Fat Chance: Beating the Odds Against Sugar, Processed Food, Obesity, and Disease* (2012, Hudson Street Press, New York). Thomas Babor and colleagues review contemporary policy approaches to the control of alcohol in their volume *Alcohol: No Ordinary Commodity: Research and Public Policy*, 2nd ed. (2010, Oxford University Press, Oxford).

POSTSCRIPT

Proceedings of the alcohol biology symposium entitled "*In Vino Veritas:* The Comparative Biology of Ethanol Consumption" can be found in the journal *Integrative and Comparative Biology* (2004, 44:267–328).

INDEX

Abstention, 40, 78–79; and longevity, 46, 47 *fig.*
Acamprosate, 95
Acetaldehyde, 41, 41 *fig.*, 44, 94, 102, 105–107
Acetate, 41
Acetic acid, 22, 25, 41 *fig.*, 120, 133
Addictions, 10, 66–67, 99, 100–101, 127, 132–134. *See also* Alcoholism
Adenosine triphosphate, 22
ADH. *See* Alcohol dehydrogenase
Africa, 77
Agoutis, 15–16
Agriculture, 7, 12, 19, 28, 64, 72, 106, 108, 118
Alcohol: and abstinence, 40, 46, 47 *fig.*, 78–79; aversion to, 26; cardiovascular benefits of, 46–47; consumption rates of, 76–82; as dietary component, 10, 78; as disinfectant, 76; energy content of, 22, 78; and food consumption, 81–82, 84–87; hormetic effects of, 44–50; industrial production of, 27, 69, 78; levels in beverages, 77; levels in blood, 37, 38, 58, 86–87, 91–93, 111, 119, 123; levels in fruit, 4, 28, 29, 30, 121; 122; and longevity, 6, 43–47, 45 *fig.*, 47 *fig.*, 125; and motor vehicles, 1, 91–93, 128; in Muslim countries, 8, 77, 129; odor plume of, 6, 17, 26, 39, 55–56, 120; physiological effects of, 79–80, 101–104; preference for, 40, 83; psychoactive effects of, 57, 70, 73–74, 79–80, 86–87, 93, 134; regulation of, 130–131; responses of lab animals to, 109–114; and sex differences, 47, 48, 78, 90; in social culture, 8, 80–84; as solvent, 76; taxes on, 132; tolerance of, 23, 41–43, 80–81, 89, 90, 110, 112, 127; vapor concentrations of, 44, 45, 45 *fig.*; withdrawal from, 89, 109, 110, 112, 127
Alcohol dehydrogenase (ADH), 5, 40, 41 *fig.*, 42, 43, 102–103, 105, 125; age of East Asian alleles, 104–106
Alcoholism: and acetaldehyde accumulation, 94; consequences of, 1, 8, 90–91; definition of, 88–90; in East Asia, 105, 107; economic impact of, 91; environmental influences on, 8, 96–97, 113; genetics of, 8, 96–99, 107–109, 111, 113; and preference for sweets, 99–100; relapse rates of, 94;

sex differences in, 8, 90, 112;
spontaneous remission of, 94;
treatments for, 8, 91–96
Aldehyde dehydrogenase (ALDH),
5, 41, 41 *fig.*, 42, 43, 94, 102–103,
105–107, 125
ALDH. *See* Aldehyde dehydrogenase
Alkaloids, 37
Amazon, 3, 4
Amylase, 73, 108
Anaerobic fermentation. *See*
Fermentation
Angiosperms. *See* Flowering plants
Animal models, 9, 109–114, 138–139
Animals Are Beautiful People, 35, 36
Antabuse. *See* Disulfiram
Antimicrobial compounds, 25, 31
Ants, 32, *plate 4*
Aperitif effect, 57, 84–85, 123
Aphids, 32
Artificial fruits, 56, 120–121
Artificial selection, 2, 28
Assam, 34
Astrocaryum standleyanum, 15, 28,
plates 2, 7
Atherosclerotic plaques, 46
Australia, 41
Australopithecines, 61

Bacteria, 2, 3, 16, 19, 23, 25, 31, 45, 48, 72,
122
Bananas, 77, 81
Barley, 71, 72, 73, 105
Barro Colorado Island, 13–14, 16, 29,
137, *plate 1*
Bats, 4, 125. *See also* Flower bats; Fruit
bats
Bears, 4, 126
Beer, 27, 83. *See also* Brewing
Beer gut, 22
Berries, 4, 104
Bertram palm, 38
Binge drinking, 48, 78, 90, 119
Bingeing, 52, 85
Birds, 3, 6, 12, 54, 59, 125

Blackbirds, 35
Blood-alcohol concentration. *See*
Alcohol: levels in blood
Blood-brain barrier, 79
Blueberries for Sal, 4
Bonobos, 137, *plate 10*
Bread, 73
Breathalyzers, 91, 115, 123
Brehm, Alfred, 117
Brewing, 7, 21, 27, 72, 73
"Burnt wine," 75
Butterflies, 36, *plate 5*

Caffeine, 100
Cage responses, 111, 112
California, 83
Calories, 18, 55, 57
Cancer, 106
Candidate genes, 98, 111–112
Cannibalism, 7
Carbohydrates. *See* Sugars
Cardiovascular risk, 46, 48, 125
Carnivores, 126
Caterpillars, 32
Catfish, 3
Cedar waxwings, 35
Cereals, 72
Chaang, 105
Cheapdate mutation, 43
Cheese, 27, 72
Chicha, 74
Chickens, 43
Chimpanzees, 4, 7, 18, 61–63, 112,
plates 8, 9
China, 5, 8, 70, 71, 75, 102, 104–105, 106,
129
Chinese monkey king, 5
Churchill, Winston, 128
Cirrhosis, 106
Cocaine, 10
Color vision, 60
Congo, 4, *plate 10*
Conifers, 14
Corn, 74
Cretaceous, 12, 22, 23

Cumbria, 35
Cycads, 14

Darwin, Charles, 116–117
Decomposition. *See* Rotting
Desserts, 86 *fig.*, 87
Diabetes, 65, 118, 132
Diagnostic and Statistical Manual, 89–90
Dickinson, Michael, 138
Dictionary of American Slang, 128
Dietary mismatch, 7, 9, 65–68, 118–119, 134
Digestive system, 3
Dinosaurs, 14
Diseases of nutritional excess, 7, 118–119
Distillation, 8, 27, 74–76
Disulfiram, 94, 131
DNA sequences, 42, 104, 120, 125
Domestication: of animals, 116; of plants, 7, 12, 19, 28, 64, 72, 106, 108, 118
Dominy, Nate, 30
Dopamine, 42, 79, 99, 100
Drosophila. *See* Fruit flies
Drosophilidae. *See* Fruit flies
Drunken monkey hypothesis, 6, 68, 115, 133, 137
Drunkenness. *See* Inebriation

East Asia, 102–104. *See also* China; Japan; Korea
Eastern Han dynasty, 75
Elephants, 34, 35
Enamel, 60
Eocene, 60
Epigenetic effects, 97
Ethanol. *See* Alcohol
Ethanology, 119
Ethyl alcohol. *See* Alcohol
Ethyl glucuronide, 38
Evolutionary medicine, 10, 65–68, 118–119
Excise taxes, 132
Extrafloral nectaries, 32, *plate 4*

Falsifiability, 115
Fats, 18, 101, 118, 119

Fecundity, 6
Feeding rates, 30
Fermentation, 2, 9, 21–22, 69–74, 121, 123–124
Fetal alcohol syndrome, 90
Figs, 16, 30, *plates 6, 8, 9, 11*. *See also* Fruits
Fingernails, 60
Fish, 3
Flight and inebriation, 37, 42
Flooded forests, 3
Flower bats, 31
Flowering plants, 3, 12, 15. *See also* Fruits
Flowers, 12, 13, 16
Foraging distances, 54
Freeze distillation, 75
"Frozen-out wine," 75
Fructose, 132. *See also* Sugars
Frugivory, 4, 13, 15–16, 52–59, 121–122; consumption rates during, 29, 36, 121; definition of, 3; among great apes, 62 *fig.*
Fruit bats, 4, 30, 37, 125
Fruit flies, 4, 24, 39–43, 113, 125, 139; and hormesis, 44–45; larvae of, 40, 44–45; reproductive fitness of, 45, 125; upwind flight of, 55, 120
Fruits: colors of, 12, 17, 54, 122, *plate 3*; competition for, 52; diversity of, 15, 83, *plate 1*; evolution of, 12–13; fermentation of, 9, 23–26, 121–122; of figs, 16; in highlands, 63; location of within forest, 53–54; odor of, 54–57; of palms, 2, 15, *plate 2*; seasonality of, 6, 16, 52–53, 126; sugar content of, 2, 17; surface-area-to-volume ratio, 31; texture, 57; unripe, 16–17. *See also* Ripening syndrome
Fungi, 3, 106, 122. *See also* Yeasts
Fusel oils, 22

Genome-wide associations, 98
Gibbons, diet of, 7, 63
Glucose, 22. *See also* Sugars

Gorillas: diet of, 7, 63, 125; in highlands, 7, 63
Grapes, 71. *See also* Wine; Wine-making
Great apes, 61. *See also* Primates
Group foraging, 53
Gun control, 130

Hamsters, 110
Hangover mutation, 43
Happyhour mutation, 43
Hawthorn, 71
Heart attacks, 46
Hemoglobin, 97, 106
Henan, 71
Hepatitis B, 106
HMS *Beagle*, 117
Hominids, 61. *See also* Primates
Hominoids, 61. *See also* Primates
Homo, 61, 62 *fig. See also* Primates
Honey, 70, 71, 77
Honeydew, 32
Hormesis, 6, 44–50, 67–68, 80, 124–125
Hornbills, 16
Humans, diet of, 7, 12, 64–68. See also *Homo*
Humidity, 4, 19, 20
Hummingbirds, 31
Hunger level, 53
Hunter-gatherer diets, 63

Indigenous peoples of North and South America, 103–104
Inebriation, 5, 38, 42–43, 80–81, 119
Inebriometer, 42
Infrared analyzer, 121–122
Insects: as agents of spore dispersal, 24; as frugivores, 6; larvae of, 3, 5, 19, 40, 44, 100. *See also* Fruit flies
Insulin, 100, 133
Inuits, 104
Iran, 71
Israel, 30

Japan, 49, 71, 81, 102, 104, 107

Keystone resources, 16
Killer yeast strains, 25. *See also* Yeasts
Kimchi, 72
Kinkajous, 15, 126
Klatsky, Art, 46
Korea, 102, 104, 107
Korine, Carmi, 30
Kumis, 105

Lactic acid, 22
Lactose intolerance, 108
Latitudinal species gradient, 15
Legal drinking age, 131
Limbic system, 99
Liqueurs, 82
Liver, 79, 100, 106, 133
London, 116
Longevity, 6, 43–47, 45 *fig.*, 47 *fig.*, 125
Long-tailed macaques, 38
Lorikeets, 37

Malaria, 106
Malaysia, 31, 129
Mammals: as experimental models, 9, 109–114; as frugivores, 3, 6, 12, 54, 59, 125; inebriation of, 42. *See also* Primates; Rodents
Marula fruit, 36
McGovern, Patrick, 71
Mead, 71
Memory, 70
Mesopotamia, 71
Metabolic syndrome, 133
Mice. *See* Rodents
Microbes, 2, 19, 20, 21
Middle Ages, 76
Migrations, 18
Milk, 108
Millet, 77
Miso, 72
Mistletoes, 30
Moderate drinking, 46–47, 49, 130
Modern Drunkard, 128
Moldova, 77
Mongolia, 105

Monkeys, 15, 62, 70, 99, 120. *See also* Primates
Monkey wine, 71
Morocco, 129
Morphine, 10
Morpho, 29
Mortality. *See* Longevity
Mutualisms, 3, 13, 26

Naltrexone, 95
National Highway Traffic Safety Administration, 128
National Institutes of Health, 129–130
Nectar, 13, 31, 37, 38
Negev, 30, 37
Nematodes, 45
Neolithic archaeology, 71–72
Neural pathways, 10, 79
New Orleans, 138
Nicotine, 10, 37, 100
Nixon, Richard, 129

Obesity, 65, 118, 132
Olingos, 126
Orangutans, 63

Paleolithic diet, 51, 117–118
Palms, 15, 30. See also *Astrocaryum standleyanum*
Palm sap, 81
Pan. *See* Bonobos; Chimpanzees
Panama, 13, 30. *See also* Barro Colorado Island
Parasitoid wasps, 44
Pasteur, Louis, 21
pH, 23, 24
Pheromones, 55
Photosynthesis, 12
Phyllostomids, 125
Pigeon breeding, 116
Pinshow, Berry, 30
Piraíba, 3
Pliny the Elder, 127
Pollination, 13, 32, 39
Polyphenols, 48

Potatoes, 81
Predators, 53, 57, 66
Primates: diet of, 6, 7, 57–60, 62 *fig.*; evolution of, 7, 59–64; laboratory responses of, 112–114; methods of fruit selection, 57, 59; olfactory sensitivity of, 56; teeth of, 60; vision of, 60
Prohibition, 128–129
Protozoans, 21
Psychoactive effects, 8–10, 57, 67, 70, 73, 79–80, 86, 93, 100, 118, 127
Pteropodidae, 125

Rainforest, 4, 6, 11–16, 18, 38, 52, 59, 63, 85, 120, 126–127, 139
Rats. *See* Rodents
"Red face" sydrome, 102
Redwings, 35
Resins, 71, 72
Restaurant profits, 82
Rhesus macaques, 112
Rice, 71, 72, 73, 106
Ripening syndrome, 2, 15, 16–18, 19, 25, 29, 57, 121–123, *plate 3*. *See also* Fruits
Road deaths, 91
Rodents, 9, 15–16, 42, 45, 56, 85, 99, 109–112
Rotting, 16, 20, 21, 25, 122
Russia, 77

Saccharomyces cerevisiae, 21. *See also* Yeasts
Sake, 49
Sánchez, Francisco, 30
Satiation, 58, 85–87
Sauerkraut, 27, 72
Seeds, 12, 14, 16, 19, 32
Sensory bias, 58, 65, 83, 87, 127
Sickle-cell anemia, 97, 106
Singapore, 30
Situational specificity of tolerance, 81
Smoking, 131
Solomon's lily, 39
Sorghum, 72, 77

Spectroradiometer, 122
Split-twin studies, 8
Standard drink, 47 *fig.*, 49
Stress response, 98, 112
Sugar cane, 73
Sugars, 22, 73, 101, 118, 119, 132–133; and alcohol consumption, 85; in fruit, 17, 18, 23, 24, 132; in liqueurs, 82
Sunbirds, 31
Supermarkets, 18, 19, 57, 65, *plate 10*
Symposium, 138

Tang dynasty, 75
Tartaric acid, 71
Temperate zone, 4, 9, 12, 14, 15, 18, 25, 28, 35, 41, 70, 110
Temperature, 4, 32, 41
Termites, 20–21
Terroir, 83
Tibet, 105
Toucans, 2, 3, 4, 16
Treeshrews, 38, 139
Tropics, 13–16, 18, 20, 33, 35, 36, 52, 126. *See also* Rainforests

United Kingdom, 35, 131
United States, 1, 69, 77, 78, 89, 90, 91, 128, 129, 130, 131, 132

Upwind flight, 39, 120
Urine production, 80
U-shaped dosage-response curve. *See* Hormesis

Vapor distillation, 75–76
Vinegar. *See* Acetic acid
Viruses, 3
Vision, 13, 54, 121

War on Drugs, 129
Wheat, 73
Wind, 55
Wine, 71, 82–84; red wine, 47, 48; white wine, 47
Wine-making, 7–8, 21, 27, 71–73, 78
World Health Organization, 77, 91

Xi'an, 129

Yeasts, 4, 18, 20, 22, 28, 121, 123–124; within fruit, 23–27, 31; within nectar, 31; spores of, 24
Yellow Mountains, 70
Yemen, 78
Yogurt, 72
Youngman, Henny, 137

Zagros Mountains, 71